MAURICE MERLEAU-PONTY'S
PHENOMENOLOGY OF PERCEPTION

**Recent Titles
in Contributions in Philosophy**

Liberty: Its Meaning and Scope
Mordecai Roshwald

Buddhist Epistemology
S. R. Bhatt and Anu Mehrotra

Partial Reason: Critical and Constructive Transformations of Ethics and Epistemology
Sally E. Talbot

Certainty as a Social Metaphor: The Social and Historical Production of Certainty in China and the West
Min Lin

Antisthenes of Athens: Setting the World Aright
Luis E. Navia

The Heidegger-Buber Controversy: The Status of the I-Thou
Haim Gordon

Aquinas in the Courtroom: Lawyers, Judges, and Judicial Conduct
Charles P. Nemeth

The Death and Resurrection of the Author?
William Irwin, editor

Beyond Subjectivism: Heidegger on Language and the Human Being
Abraham Mansbach

Nietzsche, Psychohistory, and the Birth of Christianity
Morgan Rempel

Bertrand Russell on Nuclear War, Peace, and Language: Critical and Historical Essays
Alan Schwerin, editor

Parmenides of Elea: A Verse Translation with Interpretative Essays and Commentary to the Text
Martin J. Henn

MAURICE MERLEAU-PONTY'S *PHENOMENOLOGY OF PERCEPTION*

A BASIS FOR SHARING THE EARTH

HAIM GORDON
AND SHLOMIT TAMARI

Contributions in Philosophy, Number 89

Westport, Connecticut
London

Library of Congress Cataloging-in-Publication Data

Gordon, Hayim.
  Maurice Merleau-Ponty's Phenomenology of perception : a basis for sharing the earth / Haim Gordon and Shlomit Tamari.
    p.   cm.— (Contributions in philosophy, ISSN 0084-926X ; no. 89)
  Includes bibliographical references and index.
  ISBN 0–313–32372–0 (alk. paper)
    1. Environmentalism—Philosophy.   2. Environmental responsibility.
    3. Merleau-Ponty, Maurice, 1908–1961. Phânomânologie de la perception.
  I. Tamari, Shlomit.   II. Title.   III. Series.
  GE195.G67 2004
  304.2—dc22   2003066141

British Library Cataloguing in Publication Data is available.

Copyright © 2004 by Haim Gordon and Shlomit Tamari

All rights reserved. No portion of this book may be reproduced, by any process or technique, without the express written consent of the publisher.

Library of Congress Catalog Card Number: 2003066141
ISBN: 0–313–32372–0
ISSN: 0084–926X

First published in 2004

Praeger Publishers, 88 Post Road West, Westport, CT 06881
An imprint of Greenwood Publishing Group, Inc.
www.praeger.com

Printed in the United States of America

The paper used in this book complies with the Permanent Paper Standard issued by the National Information Standards Organization (Z39.48–1984).

10 9 8 7 6 5 4 3 2 1

Contents

Introduction	1
Part One: Rejection of the Subject-Object Dualism	**11**
1 The Poverty of Empiricism	13
2 Distortions in Empiricism and in Rationalism	29
3 Insights and Problems in Gestalt Psychology	51
4 The Human Body as a Phenomenological Source	63
Part Two: Freedom and the Field of Perception	**79**
5 The Human Body and the Field of Perception	81
6 Freedom and Perception	91
7 Responsibility	105
Part Three: Sharing the Earth	**119**
8 Sharing the Earth as a Whole Person	121
9 Honest Engagement in Sharing the Earth	133
Selected Bibliography	141
Index	145

Introduction

The past four decades have seen a flourishing of discussions by philosophers, environmentalists, scientists, politicians, and laypersons on the environmental damage done to the earth by human beings. Many of these thinkers and activists have discussed the continual ruining of the natural world by human beings; they have repeatedly demanded that we human beings decide to share the earth with other natural species and not destroy the other beings that exist in the world. Some philosophers have discussed topics such as human responsibility for the world, environmental ethics, and human stewardship of the earth. Yet even these thinkers and activists have not attempted to clarify, on an ontological level, what they mean by the locution: sharing the earth with other species. We believe that many truths about the possibility of sharing of the earth with other species can be learned from Maurice Merleau-Ponty's important study, *Phenomenology of Perception*.

As part of this broad discussion, environmentalists and scientists have presented a plethora of scientific facts, findings, and conclusions on the sad state of the beings of nature and on the daily ecological deterioration of the earth. Many of their findings describe in detail how human intervention in natural processes has led to the disappearance of many thousands of living species from the face of the earth. One such book, which presents many pertinent facts and to which we return, is *The Diversity of Life*, written by Edward O. Wilson, a biologist who teaches at Harvard University.[1] Another enlightening study, which has three authors and reveals how the widespread use of chemicals in agriculture and industry ruins the hormone systems of human beings, animals, and plants is *Our Stolen Future*.[2] Yet the environmentalists and scientists who authored these books have proposed few viable or profound suggestions on how to change our basic existential relationship to the world for the better. Put differently, these works, and many other worthy works by environmentalists, have almost no relationship to philosophical thinking.

Despite the widespread concern dedicated to saving the natural environment from destruction, our rather broad survey of relevant literature has revealed three perturbing facts. The first fact is that almost none of the discussions that deal with saving the natural environment or sharing it with other species have related to the thinking on phenomenology and ontology that flourished in the twentieth century. In addition, very few of these discussions related to metaphysics, that is, to questions concerning the Being of beings. To give just one pertinent example, these discussions did not learn from Martin Heidegger that it is possible to relate to the human entity, Dasein, as Being-in-the-world. By merely comprehending Heidegger's term, we can reach a simple conclusion. Viewing the human entity as Being-in-the-world implies that ruining the world entails detracting from, and perhaps destroying, the being of all human entities which are Beings-in-the-world.

The second perturbing fact is that most of the discussions on saving the natural environment and on sharing it with other species lack a sound philosophical foundation. These discussions are not based on any ontological thinking. As this book will suggest, most participants in these discussions have little understanding of perception and of fields of perception. But our primal relationship to other beings is that we perceive them.

The third perturbing fact is that there has been virtually no profound and broad discussion of how the phenomenology of perception developed by Maurice Merleau-Ponty can contribute to our thinking on sharing the earth with other species. Our survey of relevant material has disclosed that the thinking of Merleau-Ponty has been largely ignored by environmentalists and by other noble people who are struggling to halt the destruction of our natural environment. The survey also revealed that scholars of Merleau-Ponty have not discussed in any depth the implications of his thinking for the challenge we human beings face if we wish to share the earth with other beings. Hence, this book will present some of the truths that Merleau-Ponty describes concerning perception. But it will relate to these findings primarily as a basis for a discussion on our responsibility, as human beings, to share the earth with other beings.

During his short life (1908–1961), Maurice Merleau-Ponty wrote and lectured on many topics, from phenomenology to history, and from politics to art. Yet, it is widely accepted that his study of perception, which is based on the phenomenological approach developed by Edmund Husserl, and is aptly titled, *Phenomenology of Perception*, is his most important contribution to contemporary philosophy.[3] This book is central to the ideas that we present in the following pages.

One of the central themes of *Phenomenology of Perception* is Merleau-Ponty's attack on the subject-object dualism that stands at the basis of much modern and contemporary thinking. Merleau-Ponty points out that this dualism, which in the modern period commenced with the writings of Descartes and the English empiricists, continues to serve as the basis for much philosophical and scientific thinking. From the subject-object dualism stem additional dualisms that are deemed to be central to human existence, for instance, body and soul, consciousness and world, culture and nature.

Merleau-Ponty explains that the broad acceptance of these dualisms leads to a viewing of the subject, the soul, consciousness, and culture as primary, and an emphasizing of these supposedly primary aspects of human existence. One result of this emphasis, according to Merleau-Ponty, is that the human body, the world, nature, and the other manifestations of the object that emerge in these subject-object dualisms are considered to be marginal. They are often perceived as objects that exist solely in order to serve the subject. Merleau-Ponty points to a quite disastrous result that arises from the approach that views the subject as primary: the human body, as a being that relates concretely to the world, seems to have almost vanished from philosophical discussion. Instead, the body is considered by philosophers, and even by many psychologists, to be only a means which conveys to the subject the so-called messages and impulses that come from the world. The body also performs what the subject decides to do in response to these messages and impulses.

Merleau-Ponty holds that the belief in a subject-object dualism is basically wrong, and that this belief distorts perception and human existence. Consequently, one of the central themes of *Phenomenology of Perception* is to show how the subject-object dualisms that were accepted and promoted in the philosophical thinking of the past four centuries need to be discarded in favor of a more primal and accurate understanding of human existence and perception. Merleau-Ponty addresses this challenge by first rejecting the tenets and arguments of empiricism and rationalism. He does so by carefully examining and discussing the way the human body exists in the world. He then concentrates on revealing what constitutes the phenomenon of human perception as our basic way of Being-in-the-world. One of his findings is that the human consciousness which perceives is, in the act of perception, essentially engaged in a sharing of the world.

Merleau-Ponty's goal is to consider and to elucidate human perception as a manifestation of consciousness in a person's day-to-day bodily engagements. Yet, we do not think that all of his discussions on human existence are enlightening. For instance, Merleau-Ponty's discussions of the human person as a sexual being are shallow and, at times, erroneous. One reason for his errors and shallowness is that, while discussing sex, Merleau-Ponty ignores the phenomenon of love. We shall not enlarge on such failings. Put differently, in what follows, we only discuss some of those ideas that Merleau-Ponty presents that disclose important truths. We shall not point to the few sad mistakes that appear in his enlightening presentation of perception.

We admit that our beliefs concerning human existence have probably influenced and inspired our quest to learn from Merleau-Ponty's wisdom. One of our major beliefs is that, as human beings, we are called upon daily to struggle for justice. Such a struggle should be dedicated to the principle that no person has the right to purposely destroy the freedom, the dignity, and the lives of other human beings. In addition, justice demands that no person has the right to greedily exploit and ruin the natural environment that sustains all human beings and to which we daily relate. These principles determine that each person is called upon, in his or her everyday life, to honestly share the world with other persons, whose freedom he or she should respect and support. These principles also suggest that every person is called upon to share the world with other biological species.

Given these beliefs and their immediate implications, when we discovered Merleau-Ponty's ontological insights concerning the rejection of the assumptions of empiricism and rationalism, we were excited. His ideas on the phenomenological and ontological basis of human perception, and his description of perception and fields of perception, added to our excitement. We grasped that his insights and ideas can suggest that sharing the world is ontologically basic to human perception. Merleau-Ponty also shows that it is precisely this ontological basis of sharing that the empiricists and rationalists, and many of their critics, chose to ignore. In this book, we shall frequently go beyond Merleau-Ponty's important findings. We shall develop an idea that we do not believe that he discussed. The idea is: the ontology of sharing that Merleau-Ponty describes in the act of perception has significant implications for our relationship to natural species and to other beings that exist in the world.

While studying the findings of Merleau-Ponty, we stumbled upon what seems to be a major reason why the assumptions of empiricism and rationalism are hardly ever questioned within Western society. We discovered that these assumptions are very much attuned to the dominating corporate capitalist regime. Put differently, we hold and shall repeatedly suggest in the following chapters that empiricism and rationalism are quite unable to challenge the sordid and unspiritual mode of existence that capitalism, and especially corporate capitalism, daily promotes.

It is hardly novel to state that capitalism is a regime founded upon the pursuit of greed and upon the continual exploitation of other human beings and of the riches of the earth. The ideal capitalist regards other human beings, and other living species, as mere objects or resources to be utilized for his or her own benefits. Other human beings or natural species will rarely be viewed as beings with whom the capitalist shares his or her being. The philosophical consequences are evident. These other human beings or living species may arouse in each person what empiricists call a person's sense experiences, and not much more. You will rarely—if at all—find a capitalist who will view sharing the earth with other persons or with other species as central to his or her being.

Promoters of capitalism have proffered humanity a belief that helps them partially conceal the widespread exploitation and greed that is at the foundation of their endeavors. Thus, they encourage and spread the belief that capitalism brings unlimited progress to humanity. They repeatedly declare that such unlimited progress is, by definition, Good. In many areas of the world where capitalism reigns unchallenged, this belief in the good brought by unlimited progress has become a cult. Many pundits who admire capitalism add to the cult a major principle, which is supposed to be true and hence in no need of being questioned. The principle announces that the ongoing unlimited and beneficial progress brought forth by capitalism is based upon and requires constant technological innovation. Within this atmosphere of admiration of progress, almost nobody in the political or economic realm stops and asks: Where is this unlimited progress leading us? Where is it leading the world?

A few philosophers—Jean-Paul Sartre, Hannah Arendt, Martin Buber, Martin Heidegger—did ask these questions. But their questioning has been hardly heard in the polyphony of voices that daily rave about the benefits of progress and of the wonders of technology. Consider Heidegger's essay "The Question Concerning Technology."[4] In that essay, which was written more than half a century ago, Heidegger shows that the rise of modern technology brought about a new manner of relating to the beings that exist in the world, and also to other human beings. He explains that the essence of modern technology leads us to relate to all other beings in the world, including human beings, merely as a group of resources that stand in reserve for our use. It is evident that this manner of relating to other beings alienates persons from the world that we share with those other beings. Furthermore, relating to every being as a resource to be used also promotes unending and wasteful consumption.

If Heidegger has unconcealed an important truth—and we believe that he has—capitalism coupled with the admiration of technology promote a mode of existence that cannot bring forth good. This technology-oriented capitalist mode of existence holds that the beings that exist in the world, including other persons, are there solely in order to be resources that are exploited for the needs and pleasure of some human beings and the profits of some corporations. In this mode of existence, Kant's kingdom of ends, in which each person is an end who deserves respect, has vanished. All respect for other natural species found in the world also vanishes.

Once again we can perceive where the so-called objective prism of empiricism, and the objectivity promoted by rationalism, are helpful for the technology-oriented capitalists. They are helpful because, as already intimated, empiricism and rationalism never suggest that human perception is founded upon sharing. They never relate to the human entity as a Being-in-the-world, who cares for and is responsible for the world that is central to its being.

The results of the triumph of empiricism and rationalism are sad, and often destructive. Encouraged by the princes of capitalism and their many intellectual lackeys, a technology-oriented society dedicated to the cult of progress is promoted by the media, political advisers, politicians, capitalists, and sycophantic academics. This society, which ignores ontology and embraces empiricism and rationalism, daily extols and promotes the cult of progress and the admiration of technological innovation. It ignores the fact that persons are alienated within society when they are considered to be mere resources. It also overlooks the destructive outcomes for society and for persons that result from this alienation.

Consider just a few destructive results. Persons living in the contemporary technology-oriented society learn to accept the fact that they are mere resources; hence, they dedicate themselves to finding ways to benefit from the technological innovations that are produced. They give up pursuing things that are worthy in themselves, such as wisdom, justice, love, and beauty. By such decisions, they sell their birthright for a pot of lentil soup; they continue to support the alienating cult of progress, which continues to alienate them.

Within this vicious circle, we want to again stress that the assumptions of empiricism and rationalism fit very well. These assumptions do not challenge our technology-oriented consumer society, as, say, existentialists since Kierkegaard have done. Many supporters of empiricism and rationalism seem to have no major problem with the fact that persons are viewed primarily as objects, say, as consumers or as resources. Consequently, the possibility that a belief in sharing the world with other beings will emerge from their rationalist or empiricist philosophical assumptions is almost nil.

Merleau-Ponty's insights can lead to conclusions that eradicate the vicious circle that we described. When we grasped his insights, we understood that, within a capitalist-oriented regime and Weltanschauung, there is no need to question the assumptions of empiricism or rationalism. The assumptions fit well with the goals of capitalist theorists. Together with this lack of questioning, we have witnessed the fact that many thinkers and lay people applaud both the cult of continual progress and unlimited technological innovation. They refuse to see the links of this cult to capitalist exploitation and greed, or to the ruining of nature. Hence, few of these thinkers have pointed to the dangers of accepting the so-called objective empiricist or rationalist approaches that support these evils. Fewer have pointed out that embracing the cult of progress and the worshiping of technological innovation have provided major justifications for many of the culprits who have brought about much evils in the world—including the large-scale devastation of the environment.

Part One of our book, which comprises four chapters, presents some of Merleau-Ponty's arguments for the rejection of subject-object dualism and for a renewed emphasis on the human body. Chapters 1 and 2 deal in an abbreviated manner with his rejection of empiricism and of rationalism. We believe Merleau-Ponty's arguments concerning empiricism and rationalism to be correct and well founded. We shall also suggest how his well-articulated rejection of the the subject-object dualism leads to our understanding of human perception as an engagement which is based on sharing the world.

Merleau-Ponty repeatedly admits that, in presenting his understanding of the phenomenology of perception, he learned much from the findings of the psychologists who adhered to the Gestalt school. He explained that many of the findings of the Gestalt school, and especially those findings linked to the tenet that we always perceive an image on a background, contributed significantly to his thinking on perception. Yet, as we shall briefly show, Merleau-Ponty had to also reject the thinking that serves as the foundation of Gestalt psychology. His main criticism of Gestalt psychology is that Gestalt thinking is merely an offshoot of empiricist thinking. Like empiricist thinking, Merleau-Ponty stated, the Gestalt thinkers view reality from an external perspective. Hence, the Gestalt school also denies the complexity and the validity of a free consciousness as crucial to any understanding of perception. In Chapter 3 we describe in brief the significance of the findings of Gestalt psychologists for Merleau-Ponty's ontology. We also outline the problems that Merleau-Ponty finds in Gestalt psychology. We suggest how the findings of Gestalt psychologists also support an ontology that leads to sharing the world.

The involvement and engagement of the entire human body in perception is a dominant theme in Merleau-Ponty's ontology of perception. He repeatedly points out that the subject-object dualist ontology, which prevailed for at least three centuries, has reduced the role of the human body in perception. Chapter 4 describes the human body as a phenomenological source of perception, and shows the significance of this involvement in perception for human existence and for the challenge of sharing the world with other species.

Part Two describes Merleau-Ponty's major contribution to our ontological understanding of perception. Instead of the empiricist and rationalist approaches to perception, instead of the subject-object dualism, Merleau-Ponty suggests the concept of a field of perception. According to this concept, central to the process of perception is the fact that I, as the person perceiving, create a field of perception, in which I perceive images on a background. What is unique about the field of perception is that I am free to choose it and I am bodily involved in this field. My body is part of the field of perception.

But creating a field of perception, in which I am bodily involved and take part, is a manner of sharing in the world. As Merleau-Ponty repeatedly states, the bodily creating of a field of perception breaks down the subject-object dualism. Consequently, we shall suggest that learning the manifestations of Merleau-Ponty's concept of the field can also serve as a potent source for a new manner of living a human involvement in the world. Chapter 5 discusses the human body as the originator of and participator in a field of perception.

The establishing of a field of perception by each person, in which his or her body is intimately involved, is not only a manner of relating and responding to the world into which a person is thrown at birth. It is, Merleau-Ponty emphasizes, an act of human freedom. There is a rather circular situation here, or, if you wish, a leap. Because, by establishing a field of perception within the world, human freedom comes into being. Merleau-Ponty holds that without such a field of perception in which my body is involved, and which emerges on the background of a world, there is no place or possibility for human freedom to express itself. Yet, only those who are endowed with freedom—persons—are able to bodily involve themselves in the establishing of a field of perception, on the background of a world.

Let us return to an above-mentioned problem, which clarifies the linkage between freedom and a field of perception. We have denounced the fact that, in past decades, capitalist thinkers spread a myth that there are great blessings in their quest for what they termed continual progress. This myth, which, as we already stated, was based upon much mendacity, blinded many people to the widespread destruction of the environment that this so-called blessed progress brought in its wake. The socially established field of perception supported by this myth concealed much of this environmental devastation. Hence, environmentalists have had to struggle to establish new true fields of perception in which the truths about the devastation of nature emerge and the destructive results of irresponsible progress is seen vividly. This struggle to establish a new field of perception was an act of freedom. We discuss some of the major links between freedom and perception in Chapter 6.

As the example reveals, freedom always emerges within a specific field of perception—within a worldly situation, as Merleau-Ponty and other existentialists explained. But freedom always implies responsibility. Merleau-Ponty agrees that I am already responsible for the field of perception which I choose and establish, much as I am responsible for the world which is its background. Choosing a field of perception in which progress is crucial is a choice that differs substantially from the choice of a field in which I strive to share the environment with other beings. We believe that one worthy challenge, therefore, that daily faces each person, is to establish a field of perception true to facts, in which responsibility for the earth and for other beings is prominent. Chapter 7 discuss this worthy challenge from different perspectives.

Part Three of the book will indicate some of the implications of our understanding of human freedom as central to establishing and participating in a field of perception. Chapter 8 briefly indicates how living one's freedom as a whole person is linked to the establishing of and participating in a field of perception. Chapter 9 suggests that Merleau-Ponty's thinking can lead to an honest engagement in the world. In order to be thus engaged, we must reject some of the major tenets of postmodernism—whose principles are dishonest and seem to support the sordid situation created by corporate capitalism.

Thus, these concluding chapters suggest that living as a whole person should lead to honest engagements in the world and to a responsible life in which one shares the earth with other species.

NOTES

1. Edward O. Wilson, *The Diversity of Life* (New York: W. W. Norton, 1992).
2. Theo Colborn, Dianne Dumanoski, & John Peterson Myers, *Our Stolen Future* (London: Abacus, 1997).
3. Maurice Merleau-Ponty, *Phenomenology of Perception*, trans. Colin Smith (London: Routledge & Kegan Paul, 1962).
4. Martin Heidegger, "The Question Concerning Technology" in *The Question Concerning Technology and Other Essays*, trans. William Lovitt (New York: Harper Torchbooks, 1977).

Part One

Rejection of the Subject-Object Dualism

1

The Poverty of Empiricism

> every perception is a communication or a communion.[1]
> —*Maurice Merleau-Ponty*

Whether we like it or not, the assumptions of empiricism serve as a foundation for much of our daily language; they also infect our thinking about the beings that exist in the world. These assumptions are accepted by almost all members of the scientific milieu, by the scientific establishment, by many philosophers, and by what many people call common sense. The assumptions of empiricism are so broadly accepted that they are seldom questioned. They are held to be objective and true. Even twentieth-century analytic philosophers who discussed the importance of questioning (for instance, Bertrand Russell) have shied away from questioning these assumptions.

This lack of questioning, we believe, has led to many errors in philosophy and in our understanding of human existence. It also has led to untenable ideas concerning the interactions between persons and beings in the world. Learning from Merleau-Ponty's *Phenomenology of Perception*, from which the motto for this chapter is taken, we believe the basic assumptions of empiricism to be wrong. We agree with Merleau-Ponty that "every perception is a communication or a communion." (We discuss this sentence in greater depth in Chapter 8.) Consequently, the broad acceptance of the assumptions of empiricism is very problematic.

What are these assumptions? The assumptions of empiricism hold that persons relate to the world primarily in response to their being exposed to instances of sense-experience or sense data. These sense-experiences, the empiricist explains, are conveyed to the senses by stimuli originating in objects in the world; from the senses the stimuli are transferred to the brain by the nerves, and they constitute the stuff of the mind. Furthermore, a basic assumption of the empiricist is what Merleau-Ponty and others call the "constancy hypothesis." The "constancy hypothesis" holds that there is a "point-by-point correspondence and constant connection between the stimulus and the elementary perception."[2]

With the help of examples, Merleau-Ponty convincingly shows that the constancy hypothesis is wrong. He shows that even an elementary perception, in order to be a perception, requires the meaning attributed to it by consciousness. Put otherwise, the constancy hypothesis reduces human perception to an external physical relationship between a person and an object. But this external relationship is invalid, because it fails to take the consciousness of the person who is perceiving into account, and hence, it fails to relate to the meaning that consciousness attributes to the object.

Based on their external view of perception, without the meaning that consciousness gives to the act of perceiving, the empiricists explain that quite a few of the sense-experiences that a person experiences call forth responses or actions. One of their examples of such a response is the child who puts his hand to the fire and quickly withdraws it. Such a response is a reflex, and indeed, reflexes are a favorite example of the external viewpoint adopted by empiricists. But everyone knows that there is life beyond reflexes, for instance when I watch a beautiful butterfly flittering among flowers, or listen to a piano concerto composed by Mozart.

Merleau-Ponty rejects all of the assumptions of empiricism. Like the constancy hypothesis—which he shows to be based on an ignoring of the facts of perception—he holds that all the assumptions of the empiricist, which are based on external physical relationships, do not accord with the facts of perception. A true description of perception must disregard these assumptions. However, the wisdom arising from Merleau-Ponty's rejection of empiricism is in the details.

Here are the main details of his rejection in a nutshell. Merleau-Ponty holds that the assumptions of empiricism ignore at least three major aspects of human existence: intentionality, the findings of the Gestalt school, and the full perceptual life of the human body. By ignoring these major aspects of human existence, he adds, the empiricists present us with a distorted and an impoverished view of our body, of the world, and of human life as engaged in the world. In short, the empiricists teach us little about a person's true relations to the world.

However, before explaining in greater detail why the assumptions of empiricism are faulty and impoverish human existence, we shall briefly consider the scholarly response, or rather the lack of scholarly response, to Merleau-Ponty's rejection of empiricism, and its relation to sharing the world with other beings and natural species.

Our survey of literature related to Merleau-Ponty and his critique of empiricism discovered some sad facts. We found no attempt, either by scholars who wish to prevent the devastation of nature or by lay environmentalists, to learn from Merleau-Ponty's ontology and from his critique of empiricism. We found no one who suggested that this ontology and critique may lead to human responsibility for the world, as a being who, through perception, shares the natural world with other beings. Much the same can be stated concerning Merleau-Ponty's critique of rationalism, which we discuss in Chapter 2. His critique has been largely ignored by scholars who are concerned with saving the diversity of nature and by environmentalists. Even those thinkers, among them some of the postmodernists, who criticize the foundations of empiricism and rationalism, have not discussed the implications of their critique for a genuine sharing with the beings that we daily encounter. (In Chapter 9, we return to the fact that irresponsibility for the world and its many beings, together with a basic dishonesty, characterizes many postmodern texts.)

We have encountered writers who present compelling ideas about the rampant destruction of the natural environment wreaked by human beings. We discuss a few of their fine works in later chapters. These writers stress that human beings need to act responsibly in their everyday relations with each other and with other beings of the world. Yet they do not criticize the assumptions and conclusions of empiricism. They do not seem to have recognized that the assumptions of empiricism contribute substantially to the acceptance and applauding of technology, and that the current worshiping of technological discoveries contributes to the devastation of the beings with whom we share the earth. Nor have contemporary philosophers, outside of a few, discussed the validity of the empiricist foundation underlying the so-called scientific quest for progress. Hence, many of these writers also support the prevailing capitalist cult of progress, a cult which we briefly rejected in the Introduction and at whose problems and shortcomings we hint frequently in this book.

In the Introduction we mentioned Edward O. Wilson's informative and lauded book, *The Diversity of Life*.[3] The book articulately describes the havoc and the ruin that has been wreaked upon thousands of living species in the world by the ongoing exploitative actions of the human race. It also presents examples which disclose that we human beings are irresponsible in our continual unbridled exploitation of the natural world. Yet, Wilson's book never links the prevailing ruinous attitudes that sanction rampant exploitation of the environment—attitudes that he condemns—to the leaders of corporate capitalism and their supporters who have adopted and spread these attitudes. Nor has he linked the ruinous attitudes to the empirical assumptions that establish and seemingly justify these attitudes. In a word, he never questions the assumptions of empiricism. Instead, myopically, Wilson believes that his scientific-empiricist approach, and the progress of science, can bring forth a change for the better.

Furthermore, *The Diversity of Life* admires modern technology and sees it as possibly providing the panacea for all our environmental ills. Wilson never ponders the essence of technology; he never criticizes the assumptions of the technology-oriented society in which we exist, which is linked to the widely applauded cult of progress. Furthermore, Wilson ignores the fact that these assumptions and this cult very much brought about the triumph of technology and have supported the tenet that consumerism is the sole way of bringing about benefits for human existence in the twenty-first century. Finally, it seems that Wilson never perceived that the witch's brew of falsehoods and ruinous attitudes to human existence, that stem from and rely upon empiricism, are promoted by capitalist pundits. Nor did he comprehend that this witch's brew lends much support to the irresponsible exploitation of nature and to the destruction of natural species.

The approach found in *The Diversity of Life* prevails in many—if not most—discussions concerning human relations to the environment. This prevailing situation demands that we carefully study Merleau-Ponty's critique of empiricism, as presented in his *Phenomenology of Perception*. As Merleau-Ponty noted, in this critique he learned much from the phenomenological thinking developed by Edmund Husserl.

We stated that Merleau-Ponty questions and rejects the idea of what today is called a pure sense-experience or sense-data, which constitutes one of the founding blocks of empiricism. In order to perceive the significance of this questioning, we briefly look at the way one of the founders of empiricism grasped sense experiences.

John Locke called these pure sense-experiences "distinct perceptions of things," by which he designated, say, the perceptions of yellow, white, bitter, or sweet. Consider a short citation from Locke's "An Essay Concerning Human Understanding."

First, our senses, conversant about particular sensible objects, do convey into the mind several distinct perceptions of things, according to those various ways wherein those objects do affect them; and thus we come by those ideas we have of yellow, white, heat, cold, soft, hard, bitter, sweet, and all those which we call sensible qualities; which when I say the senses convey into the mind, I mean, they from external objects convey into the mind what produces there those perceptions.[4]

Merleau-Ponty rejects John Locke's thoughts. Note that in the last three lines of the citation, Locke seems to accept the constancy hypothesis of the empiricists, which, as indicated earlier, Merleau-Ponty showed to be false.

In order to understand Merleau-Ponty's rejection of Locke's views, consider my perception of the redness of the carpet at my feet. When I look at the carpet, I perceive a carpet in which, together with its redness, its woolly shagginess and its being frayed are immediately present to me. The redness that I perceive is not pure, it is not a "distinct perception" as Locke assumed. The redness of the carpet is a woolly, shaggy, and frayed redness.

Hence, Merleau-Ponty concludes, there is no pure perception of red, as a "distinct perception" that is conveyed to me from the red carpet. In my perception, the redness is blended to the woolly, shaggy, frayed, red carpet that I perceive. In short, I perceive beings as beings, with all their complexity. I do not perceive a bundle of sense-experiences or what Locke called "distinct perceptions."

Furthermore, Merleau-Ponty explains, the findings of Gestalt psychology are in sharp contrast to the accepted assumptions of empiricism, including the assumptions of Locke quoted earlier. The Gestalt psychologists discovered and explained that we always perceive a complete image on a background. For instance, we hear a melody in the quiet of a concert hall, we do not hear a series of distinct musical notes. This finding contradicts the passage from Locke and does not provide support for empiricism. To give another example, when I look at my desk, I do not perceive the "distinct perception" of blackness and tube-likeness protruding on square-like flatness and brownness. Rather, I perceive my black plastic pen lying on the background of my brown wooden desk.

In this finding, Merleau-Ponty holds, the Gestalt psychologists have uncovered the basic condition of all perception. Perception is always a grasping of certain specific relations, in which the basic relation is that of an image on a background. The Gestalt that I perceive, these psychologists proved, already includes a contribution of my consciousness. Hence, in my perceptions, I do not "come by" certain ideas, as Locke suggested, but rather my perception contributes to the constitution of the images that I encounter on a background. Later in his study of perception, Merleau-Ponty points out that, unfortunately, the Gestalt psychologists did not comprehend the sweeping and enlightening implications of their findings. In Chapter 3, we present in detail some of the significant findings of the Gestalt school, and return to this problem.

In contrast to the empiricist belief, it is not elements of supposed sense data, say, redness and shagginess, that I somehow blend in order to conclude that I perceive a red shaggy carpet. Rather, I always initially perceive the carpet as an image, with its redness and frayed shagginess blended together, lying on the floor of the room which is the background. In order to only perceive, say, the redness of the carpet, I must abstract the redness from the carpet—which is a feat that requires reflection.

Merleau-Ponty describes a major consequence of this basic condition of perception that was discovered by the Gestalt psychologists. He holds that, without my perceiving an image on a background, there is no possibility of my comprehending whatever I may encounter. Thus, in the act of perception, my consciousness intentionally establishes a unique yet shared field of perception in which images may emerge on a background. Each field of perception is unique, Merleau-Ponty explains, because every person has an individual style in establishing his or her field. Each field of perception is shared with other persons who are engaged in the same section of the world at the time that I am engaged in that section of the world. These persons encounter many of the phenomena which I encounter. Each person listening to a concert establishes his or her unique auditory field which is shared with other people sitting in the concert hall.

This unique yet shared field of perception that perception establishes, and within which it relates to images on a background, is the basis of the world in which the images exist. We again mention a concept from Heidegger—Being-in-the-world—which will accompany many ideas in this book, and which Merleau-Ponty generally accepted. Learning from Heidegger's concept, we say that every moment perception contributes to establishing each person as a Being-in-the-world. Through this primary relation of perception, I find myself not a passive being who, as Locke suggested, "comes by" certain specific "sensible qualities," say, yellow, bitter, or sweet. Rather, I am a Being-in-the-world whose intentional perception establishes a field of perception. I am a person with a body, who perceives images on a background, and who shares with other beings the world that my body and my perceptions engage in and establish.

Merleau-Ponty intimates that many philosophers ignored the workings of my consciousness and the field of perception that it establishes. One reason may be that they did not take into account that my consciousness is intentional—it is usually directed toward a specific object. Every moment I bodily live this intentionality. I am looking for my pen on the table. While seeking the pen, I do not direct myself to the papers, the books, and the stapler that are strewn on the table. My eyes and my hands are solely directed toward finding the pen. Merleau-Ponty repeatedly emphasizes that I do not perceive my consciousness as being directed to an object. Rather I live my being directed to an object.

The basic fact that founds perception, Merleau-Ponty explains, is that my consciousness is always directed to an object, which it also transcends. In this act of transcending, I grasp the object as a totality. This totality is always more than the sum of its components which I may perceive. The carpet that I perceive is a totality on the floor of the room. It is more than its redness, shagginess, and other perceptible qualities. Parts of it exist also under the legs of the chair and the table. It has an underside that I do not perceive. My directing myself toward an object and my transcending it towards a totality is pre-objective and pre-reflective, in the sense that it has nothing to do with the supposedly objective world established by reflection. We return to this objective world.

The distinction between the pre-objective realm and the objective realm is not developed in the classical empiricist theories. They primarily rely on the objective realm which is developed by reflection. Put differently, the human consciousness's pre-objective pre-reflective relationships to the objects in the world are pretty much ignored by the propounders of empiricist doctrines. But pre-objective experiences constitute most of every person's daily bodily life. When you play tennis, or pick up the fork to eat a piece of melon, or tie your shoe, or brush away a fly, you are acting pre-objectively. Hence, by ignoring the human life that occurs in the pre-objective realm, empiricists further impoverish their comprehension of human existence.

Why do the empiricists stress sense-experiences, or sense data? Why does John Locke write about "distinct perceptions?" We shall proffer one possible answer. Frequently in the empiricist's thinking about the world—and in his or her abstracting one's perception of an object—instances of "distinct perceptions" of qualities seemingly appear and supposedly help the empiricist to understand the perception of the objects. For instance, if asked about the color of my carpet, I will say that it is red. From that statement, on the basis of reflection, I may conclude that I know the color of my carpet because of the "distinct perception" of red that my eyes perceived when I looked at the carpet. But this manner of thinking changes the order of my relation to beings in the world. The true order is that perception is primary, it is a going out to the object while establishing a field of perception in which that object is found. Reflection on the object is always secondary. Making a "distinct perception," that appears only in reflection, into a primary manner of relating may be called the original sin of empiricism.

We again state an important result of Merleau-Ponty's inquiry. Redness will never appear alone in my everyday interaction with the objects in the world. The redness of the rug at my feet will always be that of this specific shaggy frayed carpet. The redness of the buses in London will always be a metallic redness. Consequently, when I believe in sense-data or "distinct perceptions," I believe in an abstraction.

Does not much of the prevailing discussion of a person's relation to the environment contain similar abstractions? For instance, is not the word "sustainability" as a so-called way of saving the environment also a mere abstraction? The fact that the term "sustainability" recently became popular in many discussions about human relationships to the beings in the world (including in the title of a world conference on the environment that was held in August 2002 in Johannesburg, South Africa) does not make it less abstract. Perhaps the popularity of "sustainability" is a result of its being an abstract concept. We return to some of the problems of "sustainability" later in this chapter.

How will the empiricist respond to the critique stemming from Merleau-Ponty's ontology? We suspect that some empiricists will say that they are merely presenting a theory that is supported by scientific evidence. Implicitly, they will be using the constancy hypothesis. They will state: Certain wave lengths of color from the carpet reach my retina and affect it. This impression is transferred by my nerves to my brain, which by some, as yet unclear, processes leads me to reach the conclusion that the carpet is red.

We have already mentioned that, in response to this argument, Merleau-Ponty repeatedly points out that this scientific evidence reveals only an external view of what occurs in the process of perception. Hence the empiricist views the human entity as an essentially passive being, which it is not. Moreover, the empiricist's so-called scientific view ignores the centrality of the human body, of consciousness, and of intentionality in all perception. The view also ignores the findings of Gestalt psychologists.

My consciousness, Merleau-Ponty states, which is incarnate in my body, determines what I choose to see, which image is chosen to stand out on a specific background, and even which reflexes and feelings I will choose and endow with significance. By these everyday decisions it is precisely human consciousness that establishes the significance of the other beings that I encounter—in my immediate environment and in the world.

Merleau-Ponty adds that very early in life I discover that for this world to be a world, and to have meaning, my consciousness must be firmly wedded to the actions and to the impressions of my body. I discover that the intentionality of my consciousness, the actions and perceptions of my body, and the choices of my being, together with specific historical facts anchor me in a section of the world. My consciousness gives meaning to my doings, which include my perceptions of the beings in the world that I encounter in this section of the world in which I exist. On the pre-reflective level, I already recognize that some of the beings that exist in this part of the world affect my body, my being, and my consciousness.

As a child, and also later in life, I repeatedly learn that, often, the beings that I encounter, which include other persons, share the world with me. I learn that they are beings that I must take into account in making my choices and in determining my behavior. Later, we would add, I may learn an important principle underlying these findings: For the world that my consciousness finds and establishes to be significant and worthy, I must share my self, at least partially, with other human beings who have bodies and are endowed with consciousness; I must also share my being with the other beings that I encounter as existing in the world. This principle, that I learn through interactions of my consciousness, and all of these findings concerning the human body, consciousness, and intentionality—including the significance of sharing my being with other beings for the establishing of a world—nothing of all this is found in empiricist thinking. From this perspective, despite its supposed closeness to scientific inquiry, empiricism seems to exist in a vacuum.

Empiricist thinking and the external scientific approach which accords with it, impoverish human existence in another manner. Since these approaches ignore human consciousness and the human body as central to establishing a world, the manner by which they relate to the world, to nature, and to the environment is frequently barren and sterile. In presenting the case for empiricism, the empiricist takes no account of what Merleau-Ponty calls "the cultural world and the human world." Merleau-Ponty points out that ignoring "the cultural world and the human world" conceals much of human daily existence, and thus narrows every person's Being-in-the-world. He writes that it is

> desirable to point out everything that is made incomprehensible by empiricist construction and all the basic phenomena which they conceal. They hide from us in the first place "the cultural world" or "human world" in which nevertheless almost our whole life is led. For most of us, Nature is no more than a vague and remote entity, overlaid by cities, roads, houses and above all by the presence of other people. Now, for empiricism, "cultural" objects and faces owe their distinctive form, their magic power to transference and projection of memory, so that only by accident has the human world any meaning. There is nothing in the appearance of a landscape, an object, or a body whereby it is predestined to look "gay" or "sad," "lively" or "dreary," "elegant" or "coarse."[5]

Thus, the broadly accepted empiricist and scientific approaches and concepts often narrow our Being-in-the-world, and make many of our relations to the world incomprehensible. The beings in the world that the empiricist describes have no character, no style, no harmony or disharmony, no beauty or ugliness—we could go on. But the world in which we are engaged is a cultural human world and not only a world of sense-data or "distinct perceptions" and "sensible qualities." By implicitly suggesting that human beings daily exist in a world that is divorced from all culture, the empiricist spreads false assumptions and impoverishes human existence.

Moreover, the previous citation from Merleau-Ponty suggests that empiricist approaches and concepts frequently alienate persons from nature and from the world that we daily share with others. As already indicated, the concepts and ideas that are central to what is called the scientific endeavor and the capitalist cult of progress will often support this alienation. Thus the physical or biological sciences, which are part of our human culture, are based on the empiricist approach whose assumptions ignore the existence of human culture, especially in our daily relations to the beings in the world. Unfortunately, few empiricists comprehend this paradox that is at the basis of their approach.

We want to present an additional example of how the empiricist approach impoverishes the world in which we live and makes the beings of nature quite irrelevant. For the empiricist a rose is merely a bundle of sense-data or "distinct perceptions." The empiricist agrees that a specific rose may arouse in the person contemplating the rose memories of other such roses which were also bundles of sense-data. These memories may include incidents in which roses played an important role, say in a courting or a marriage. But in all these incidents, the rose has nothing unique about it that transcends the sense-data that supposedly constitute it. It cannot lead to our understanding additional aspects of human existence, such as the fact that the rose may be considered beautiful, or that it may signify beauty, or generosity, or love, or joy at sharing beauty with another person. For the empiricist, the rose is a thing, like many other things that are found in the world.

The empiricist's way of relating to the rose impoverishes our lives and ignores the spiritual dimension of human existence. Indeed, for the empiricist there can be no major truth disclosed in the famous statement of Gertrude Stein: "A rose is a rose is a rose."[6] To be true to his or her creed the empiricist must argue that Gertrude Stein's statement is merely a tautology. But Stein's statement is not a mere tautology. "A rose is a rose is a rose" discloses a major truth about human existence, about human culture, and about the links between our appreciation of natural beauty and our ongoing quest to unconceal truth and search for wisdom. Discussing this unconcealed truth and showing its significance in our daily existence would take us too far astray. Yet, we believe that Stein's seemingly tautological statement deserves to have become a part of Western culture—as has happened.

To summarize this point, look again at Merleau-Ponty's major point in the earlier citation. The impoverishing of human existence to which he points is a direct result of the empiricist's divorcing nature from culture. Hence, in the world described by the empiricist, natural beings do not have the possibility of being dreary, lively, beautiful, or elegant. Indeed, for the empiricists there is no possibility for a person to relate to beings in nature with fear, wonder, awe, delight, anxiety, or joy. One sad conclusion immediately emerges. By holding that scientific evidence is at the basis of his or her approach, and by ignoring that human consciousness establishes a world in which human beings exist, the empiricist is establishing an impoverished and very boring world. We would add that he or she is also establishing a relationship to the world that will not lead to a worthy enhancing culture, to spiritual existence, or to genuine sharing.

We have already stated that we agree with Merleau-Ponty's profound critique of empiricism, from which we have presented only a few elements.

Someone may still ask: Why is this critique of empiricism important if we wish to establish a basis for a genuine sharing of the world with other beings and natural species?

One reason is that even the few elements that we presented from Merleau-Ponty's critique show that the external relations of empiricism cannot account for my perceptions in the world and cannot contribute much to elucidating my Being-in-the-world. His critique clearly shows that, in contrast to the assumptions of the empiricist, my consciousness and daily bodily perceptions establish the world while being affected by it and emerged in it. Merleau-Ponty states: "We are caught up in the world and we do not succeed in extricating ourselves from it in order to achieve consciousness of the world."[7]

Let us be specific. My relationship to other beings in the world is not external. It is not like the relationship between the different balls that lie on a billiard table. Nor is my relationship to other beings in the world external yet vital, like the relationship between the rays of the sun and a green leaf which turns toward the rays. Put succinctly, because all human beings are endowed with consciousness, my relationship to the beings that I encounter never can be fully reduced to chemical or physical causes and laws. Never! But external relations between the different beings in the world are the only relations that interest the empiricist. Moreover, as we briefly showed, the empiricist wrongly holds that the external relations that he or she discusses are the only relationships that disclose what really exists in the world. As mentioned, in order to hold this position, the empiricist adopts a position that ignores human consciousness, human culture, and distorts perception.

Because there is no genuine sharing between the balls on the billiard table, or between the leaf and the ray of the sun, the empiricist cannot even contemplate an ontology that reveals the possibility of a relationship of sharing among beings that exist in the world. Need we add that the world that the empiricist constructs, a world that ignores the possibility of sharing, is a wasteland of human relations? Is it not clear that, in such a wasteland, respect for other persons and for other species that exist in the world can only rarely emerge?

Sharing is never merely a pursuing of joint interests. Genuine sharing requires respect and responsibility for the specific being whom I encounter. Nothing of such sharing, respect, and responsibility can be found in the accepted empiricist theories. Despite the reign of empiricism in science and in much common sense, there are examples of sharing in our cultural heritage. For instance, you can find a few examples of sharing, including sharing with beings of nature, in the writings of some of the world's great authors.

In some of his writings, Ernest Hemingway has shown that sharing is possible even in my relations to a being that I am hunting. Santiago, the old man in Hemingway's *The Old Man and the Sea,* shares the ocean and his being with the huge fish that he has caught and has decided that he must kill. He relates to the fish with respect and with responsibility for its being. During his three-day struggle to catch it, he speaks to the fish. From his speaking we grasp that, for Santiago, the great fish is his partner in this world, a true brother. Listen to Santiago.

"The fish is my friend too," he [Santiago] said aloud. "I have never seen or heard of such a fish. But I must kill him. I am glad we do not have to try to kill the stars."
  Imagine if each day a man must try to kill the moon, he thought. The moon runs away. But imagine if a man each day should have to try to kill the sun? We were born lucky, he thought.
  Then he was sorry for the great fish that had nothing to eat and his determination to kill him never relaxed in his sorrow for him. How many people will he feed, he thought. But are they worthy to eat him. No, of course not. There is no one worthy of eating him from the manner of his behavior and his great dignity.
  I do not understand these things, he thought. But it is good that we do not have to try to kill the sun or the moon or the stars. It is enough to live on the sea and kill our true brothers.[8]

We submit that Hemingway is one of a few twentieth-century authors who describe in vivid detail the human possibility of establishing relations of sharing with the non-human beings that exist in the world. Hemingway's writings that describe such sharing also show that sharing is a worthy way of existing in the world. But for sharing to come into being, persons must abandon the assumptions of empiricism and the external relations that these assumptions promote.

Learning from Hemingway, we believe that sharing relations will hopefully lead human beings to respect and to strive to preserve the other species and beings that exist in the world, with all their uniqueness and grandeur.

We are now prepared to fully answer the question: Why is Merleau-Ponty's criticism of empiricism significant if I wish to establish relations of sharing with other beings in the world? In presenting a full answer, we shall repeat some previous statements.

As stated, a major reason is that Merleau-Ponty's critique shows that the empiricists are basically wrong in viewing human perception as an external relation of the human body to objects in the world. External relations, like those that may exist between the balls on a billiard table, or between rays of the sun and the leaf of a plant, are never relations of sharing. Hence, as quite a few existential thinkers have shown, any attempt to base perception and knowledge on external relations alienates and demeans human existence, in addition to being wrong. Furthermore, if I wrongly view all relations that exist as external—such as exist between the balls on a billiard table—there is no rhyme or reason to discuss the manner by which I share the world with other beings, and the consequences of this sharing.

We have also indicated that the empiricists adopted their mistaken assumptions because they excluded the activity of the body and of consciousness from their discussions. Most significant is the fact that they excluded the pre-objective pre-reflective consciousness and each person's daily bodily encounter with the world. Consciousness and the actions of the body, however, are central to all perception. And perception itself, Merleau-Ponty stresses, is a communication and a communion, that is, an establishing of a shared world.

The situation should now be clear. Merleau-Ponty's thinking shows that if we comprehend perception as it exists, we will understand that no world exists or can be established on the basis of external relations—without consciousness, and without my bodily manner of perception while sharing the world. We will also comprehend that no image will appear on a background without consciousness establishing the field of perception on which the image may appear. In everyday life, you cannot separate what consciousness, intentionality, and perception contribute to grasping a specific being, say the rain pattering on my window, from that specific being that I encounter, that is, the rain. I share and establish the world that I inhabit by and through my everyday bodily perceptions which are intentional and pre-reflective.

Here we have unconcealed a truth that suggests that I should relate to the beings in the world with respect and with responsibility. The truth is that my perception constantly establishes what must be called, from an ontological perspective, a shared world. Every moment during which I am awake, my body shares my field of perception with the beings that I perceive and with whom I interact. Within this shared perceived field I establish the world while my bodily being partakes in it. This shared perceived world is my world. It is the only world that I have and inhabit.

Unfortunately, the ontological fundamentality of perception, as establishing and sharing the world, is overlooked by all those environmentalists who start from what they call the scientific perspective. By starting from this perspective they already include the flaws of empiricism in their approach. Many of these environmentalists announce that they wish to preserve nature. Some of them state that they will do their utmost to support and sustain many of the natural species that exist in the world. Many of them also struggle, with courage and nobility, to protect known endangered species. (Many endangered species, say, of beetles or algae, are unknown.)

Furthermore, a large group of environmentalists advocate adopting an approach to natural beings that they call sustainability of the natural environment. We believe this approach to be supercilious, and will soon explain. However, here we want to stress that all of these environmentalists ignore the fact that in all their pronouncements and suggestions, in which they start from the scientific perspective, they base their arguments and suggestions upon the foundations of empiricist theory and thinking. Consequently, the foundation of their thinking, which is based upon empiricism, clashes with their announced goals which should lead to sharing the world with other beings and species. Perhaps the saddest aspect of this situation is that many of the environmentalists do not perceive this clash.

Yes, even those environmentalists who advocate sustainability of the environment have an ontological problem. They overlook the fact that their understanding of sustainability is an abstraction based on external relations, which overlooks human consciousness and the phenomenology of perception. By their overlooking of the ontological foundation of their views, they do not see that they are establishing these views upon the flawed assumptions of empiricism.

In addition, there exists what may be termed an ethical problem with the concept of sustainability. We believe it also to be a problem of superciliousness. Sustainability of the environment is a form of domination—perhaps it is a more gentle form of domination, if we are to believe the adherents of this school of thought—but it is still a domination of the beings of nature by human beings. The promoter of sustainability superciliously believes that we human beings, who are alive today, can and should determine which beings and species that currently exist in the world—say, which beetles, or weeds, or fish, or algae—should continue to exist freely in their natural abode, in order that they be sustained for future generations. Implicit in the term "sustainability" is the belief that we human beings can also decide which of these beings and natural species may be allowed to vanish from the face of the earth. The advocates of sustainability never ask: Who permits us human beings to make such grave decisions? They also never ask: Why have we created a social-economic regime which brings many thousands of natural species to extinction? Should we not struggle to change the regime?

Finally, the so-called scientific perspective is also problematic. We cannot here prove, but we accept some of the profound and quite disturbing ideas of Nietzsche and Heidegger concerning the scientific endeavor. These thinkers held and endeavored to show that scientific thinking is often dictated by technological thinking and by the essence of technology. And modern technology, as is well known, is based upon an external mechanical alienated relation among beings. This external relationship, Nietzsche and Heidegger repeatedly announced, degrades human existence by totally ignoring the human ability and possibility to engage in magnificent endeavors, such as to think, to love, to seek wisdom, to write poetry, and to create and appreciate beauty.

We believe that there is much truth in Nietzsche's and Heidegger's insights and arguments that the flourishing of technology will frequently emasculate human grandeur and thinking. We agree with the idea that engaging in science is frequently linked to the human attempt to dominate and rule nature, and the world. Need we add that domination and ruling are rarely linked to sharing?

To recapitulate, by excluding the workings of consciousness from their discussion of perception, and by blatantly ignoring the ontology of perception, the empiricists have started from untenable and false assumptions. These erroneous assumptions have led them to view the world in which we are daily engaged as a mechanical, distorted, bleak, and culturally impoverished world. Merleau-Ponty painstakingly reveals that the assumptions of empiricism ignore the truth of the phenomenon of perception, which is daily expressed in language. Perception, he shows, is embedded in the realm of human culture that was established by past and present generations. Divorcing perception from culture, as the empiricists have done, leads to superficiality and to a host of errors.

Our brief critique of empiricism, which learns from some of the ideas of Merleau-Ponty, reveals that the world described by the empiricists is not a world of beauty, of wonder, or of sharing. Their description of the world hardly includes the possibility of responsibility for our natural environment and for the beings that we may encounter in the world. In that bleak impoverished world that emerges from the assumptions of empiricism, there is no meaning to Santiago's speaking as a brother to the great fish that he must kill. In that mechanistic and barren world of the empiricists, Gertrude Stein merely presented a tautology when she wrote: "A rose is a rose is a rose."

Consequently, we can already suggest that studying the wisdom of Merleau-Ponty, including his critique and rejection of empiricism, can lead to a truer vision of the world and of the place of human beings in it. We shall continue to show in the following chapters that the truths and vision emerging from his thinking can enrich and enhance our own being and can lead to greater responsibility for the world that we share with other species and beings.

NOTES

1. See Maurice Merleau-Ponty, *Phenomenology of Perception*, trans. Colin Smith (London: Routlege & Kegan Paul, 1962), 320.
2. Ibid., 7.

3. Edward O. Wilson, *The Diversity of Life* (New York: W. W. Norton, 1992).

4. John Locke, "An Essay Concerning Human Understanding," in Edwin A. Burtt (ed), *The English Philosophers from Bacon to Mill* (New York: The Modern Library, 1939), 248–49.

5. See Merleau-Ponty, *Phenomenology of Perception*, 23.

6. Gertrude Stein, *The Autobiography of Alice B. Toklas* (Middlesex, England: Penguin Books, 1966), 150–51.

7. See Merleau-Ponty, *Phenomenology of Perception*, 5.

8. Ernest Hemingway, *The Old Man and the Sea* (New York: Simon and Schuster, 1995) 75.

2

Distortions in Empiricism and in Rationalism

We have already indicated that Merleau-Ponty's critique of empiricism does not mean that he accepts the principles that are the basis of rationalism. Rather, he argues, both rationalism and empiricism cannot provide a factual basis for what happens in human perception. History discloses, however, that many rationalist philosophers, for instance Descartes, and many empiricist philosophers, such as Locke, described perception in their writings. Hence, Merleau-Ponty understands that he must explain how the descriptions of both the empiricist and the rationalist distort the facts of human perception.

Learning from Merleau-Ponty, in this chapter, we dedicate much space to a critique of rationalism and the implications of this critique for our relationship to nature and to other beings. Yet, we also discuss some of the distortions that are embraced by both the empiricist and the rationalist, distortions that complement those described in Chapter 1. Merleau-Ponty shows that the distortions of the rationalist and empiricist are not always contradictory. Consider the relation of the empiricist and the rationalist to the so-called objective world.

When discussing perception, both the empiricist and the rationalist begin by considering what they call the objective world. The objective world is frequently the object of their analysis. They are confident in relying on the objective world because it is the world addressed by the scientific endeavor. But, as any person who takes the time to examine the ontological status of the objective world will discover, the objective world is not primary in human perception. It appears later both in time and in meaning—for one reason. Reflection is needed for the objective world to appear to a person's consciousness, and reflection can only come into being after a person has had pre-objective pre-reflective perceptual experiences in the world. Put succinctly, all reflection is always based on pre-reflective experiences.

Unfortunately, both the empiricist and the rationalist pretty much ignore the activities of the pre-reflective consciousness. Because they begin with the objective world, Merleau-Ponty holds, neither the empiricist nor the rationalist is capable of "expressing the peculiar way in which perceptual consciousness constitutes its object."[1] We have already referred to the immediate results of this unwarranted ignoring of a person's pre-reflective perceptions and life. Both the empiricist and the rationalist adopt an abstract and erroneous approach to the phenomenon of perception, to phenomenology, and to human existence.

Put differently, by basing their descriptions and explanations upon the objective world, which is a world constituted by reflection, both the rationalist and the empiricist embrace abstractions and distance themselves from perception as it actually occurs. Such a distancing of oneself from perception as it occurs is wrong, and has grave consequences. Hence, in this book we repeatedly indicate that the accepting of so-called objective descriptions and explanations of the world as valid, and as encompassing all perception, is erroneous. We also show that this mistaken approach may often encourage a person to relate to the world and to other persons and species in the world in alienating and exploiting manners.

Merleau-Ponty presents a telling example of the problems that both the empiricist and the rationalist face—problems which arise from their skewed understanding of perception. In the example, Merleau-Ponty briefly surveys the history of the concept of Attention.

Merleau-Ponty explains that the empiricist cannot account for what we call "paying attention to something" since he or she presents perception only on the basis of external links and relations. Consider a simple example of my attending a recital of Maria Pires playing a Mozart piano sonata. What will happen if, during the concert, I look at her hands playing the piano, yet, I listen passively and do not pay attention to the music or to her hands—if, say, during the adagio, I reflect upon Beethoven's deafness? The answer is not difficult. According to the empiricist, the same sense-data will be registered upon my retina and on my eardrums as when I do pay attention to her exhilarating performance. Consequently, Merleau-Ponty holds, for the empiricist who describes perception solely by external relations, there can be no way to account for the difference in my perceptions when I pay attention to something or do not pay attention to something—say, to Maria Pires playing a Mozart piano sonata. But such a difference exists.

The rationalist, or, as Merleau-Ponty often terms him or her, the intellectualist, has different problems. According to the intellectualist, almost all my perceptions are clear and distinct. Those that are not clear and distinct require that I merely pay attention to them. Hence in perceiving, I always pay attention to something and there is not a need of the locution "paying attention." But this view of clear and distinct perception does not take into account a simple truth. There are moments when I pay full attention to a phenomenon, such as a recital of a Mozart sonata, and there are always many moments when my mind wanders or my attention is partial. Yet in both instances, I hear the music clearly and distinctly. Hence, attention cannot be identified as clear and distinct perception.

Merleau-Ponty adds to this critique the fact which has been researched and disclosed by the Gestalt thinkers—that I always perceive a form or a totality on a background. Note that the background, which is perceived as background, but to which I may hardly pay attention, is not necessarily clear or distinct. Usually I do not pay attention to the background of the event to which I direct myself. Thus, I may hardly be interested in the lighting or the stage curtains when attending a concert of Maria Pires playing the piano.

Merleau-Ponty also blames the rationalist for presenting a theory of perception that eradicates learning from the spectrum of human possibilities. The reason is simple. If all perception is clear and distinct, I need not seek to learn more about any item, since my consciousness already has perceived it clearly and distinctly.

Put differently, Merleau-Ponty criticizes rationalism on two major points. First, the intellectualist's view of consciousness is seemingly too rich, in that, on an abstract level, it has all the answers. Hence, the rationalist encounters many difficulties when he or she endeavors to describe how consciousness relates to a specific phenomenon. Specifically, the intellectualist cannot explain how I perceive that a phenomenon includes evident and yet concealed mystery. For instance, the beauty of a well played Mozart piano sonata includes an evident and concealed mystery that is perceived by the listener but can never fully be presented rationally. Put succinctly, beauty can be perceived but never rationally explained. Nor can the intellectualist explain why a certain phenomenon may appeal to or demand a person's attention.

Second, for the intellectualist, since my consciousness is the source of my thinking clearly and distinctly and my perceiving clearly and distinctly, I do not have to go beyond consciousness and thinking. The act of learning through acts of consciousness and by bodily engagement in the world becomes superfluous. But learning through acts of consciousness and bodily engagement in the world occurs. Think of the acts of consciousness while learning to speak a foreign language, or the bodily engagements and acts of consciousness while learning to play tennis or billiards.

We have stumbled upon a major truth concerning empiricism and rationalism. Merleau-Ponty argues that the basic misunderstanding of the phenomenon of attention is a misunderstanding that characterizes both empiricist and rationalist thinking. This misunderstanding reveals that both the empiricist and the rationalist cannot grasp human consciousness when it engages in the act of learning, which includes learning by bodily engagement. Nor can it grasp human consciousness as appreciating beauty or experiencing love, which, of course, includes perceptions.

Put bluntly, the rationalist and the empiricist cannot answer complex questions concerning learning and the perception of beauty or the experiencing of love. For instance they cannot address or respond to the question: How and when do I grasp or learn that Maria Pires plays the Mozart piano sonata beautifully? Nor can they even approach the question: How do I learn or perceive that a specific musical composition, say Verdi's *Aida*, is beautiful? Similar questions that the empiricist and rationalist cannot answer can be asked about love and other personal experiences.

Yes, because they have nothing to say about learning or about the beautiful, or about other experiences that include a mystery, neither the empiricist nor the rationalist can even pose the question: Why is this Mozart sonata beautiful and worth learning to play? For the empiricist, Maria Pires's playing is merely the creating of sound waves which reach my consciousness and supposedly arouse pleasure in my mind. For the rationalist, my attending a concert during which Maria Pires plays a Mozart piano sonata, merely results in my having a clear and distinct perception of the melody, of its tenor and harmony. And this clear distinct perception, the rationalist would admit, may give me pleasure. In both of their explanations, beauty has vanished!

On the basis of Merleau-Ponty's arguments, we hold that, because of their distorted understanding of perception, both empiricism and rationalism lead to a vacuous understanding of large sections of reality. Such a vacuous understanding of reality is sad and debilitating. As we shall repeatedly show in this book, it can hardly encourage people to relate to other species that exist in the world as worthy of sharing the world with human beings. In order to partially disclose the vacuity that arises from the empiricist and rationalist approaches to perception, we briefly digress and look at a literary example by a writer who did write about musical beauty—Marcel Proust.

It is evident that, if he were faithful to either the empiricist or the rationalist philosophy of perception, Marcel Proust could never have written *Remembrance of Things Past*.[2] He could never have vividly and profoundly portrayed the life of Marcel, the narrator of his masterpiece. Nor could Proust have described, in exquisite detail, Marcel's perceptions and bodily engagements and interactions in the world that he encounters. In the novel, Proust also has Marcel describe his reflections and the so-called objective world in which he exists. But Proust repeatedly reveals that, for Marcel, reflections and the so-called objective world are not primary. Only perceptions and bodily engagements and interactions in the world are primary.

Put differently, Proust's detailed descriptions of Marcel's reflections emerge on the basis of his pre-reflective perceptions and actions, and his bodily existence and responses. These perceptions and actions reveal how Marcel, like all persons, engages in perceiving the world and bodily relates to the world. It is therefore not difficult to discern that Proust's descriptions of human existence and perception transcend any reality that can be described by the intellectualist or the empiricist. Thanks to this transcending, Proust was able to present many insights concerning human existence, and especially the relation persons have to time. These insights include many words of wisdom, to which we can only refer. For instance, Proust describes how a person, say Swann or Marcel, learns to perceive and discern the beauty of a musical composition—in the novel it is a sonata by Vintuel. He also shows how persons learn to see the beauty of a painting—in the novel, the painter of beautiful paintings is Elstir.

Someone may still ask: What exactly in Proust's descriptions of perception and the human mind is enlightening? Here we recall only one relevant and well-known example.

Consider the description of twenty-year-old Marcel returning home after a walk and tasting a crumb of the madeleine soaked in his mother's cup of tea, in her decoction of lime flowers. Marcel's bodily response to the taste of the madeleine, and the bodily memory that it arouses, amazes him. The taste brings back warmth in his body and a host of childhood memories, and with them an entire village, Combray, and its surroundings. Listen to Marcel summarily describing the memories which arose in him in the moment of tasting the madeleine. Note that these memories have nothing to do with reflection.

in that moment all the flowers in our garden and in M. Swann's park, and the water-lilies on the Vivonne and the good folk of the village and their little dwellings and the parish church and the whole of Combray and of its surrounding, taking their shapes and growing solid, sprang into being, town and gardens alike, from my cup of tea.[3]

It is evident that neither the empiricist nor the rationalist can explain, on their grounds, how Marcel's tasting of the crumb of madeleine soaked in lime tea brought into being in his memory and daily existence all of Combray and its surroundings. For a simple reason. Their theories balk at the fact that our bodily perceptions are linked to and inhabited by memories, and can bring forth distinct memories. But Proust repeatedly shows that such is the case. And any person who consults his or her own perceptions and memories will agree with Proust's insight.

The cited example is one among many similar moments that are described in *Remembrance of Things Past*. In these described moments, Proust repeatedly describes major truths about the ways that Marcel's body and his pre-reflective consciousness relate to the world and give meaning to simple daily perceptions. These descriptions of truths about perception reveal that perception is not based on external relations, and that it is primary, in both time and experience, to reflective thinking. Proust's descriptions also disclose our bodily relations to time, to beauty, to memory, to the obsessions that accompany a failed love, and to many other relations and interactions of a lived life.

As Proust presented them, a person's bodily relations and perceptions intermingle with memories, and also with perceptions of beauty or ugliness. Bodily engagements and perceptions also situate a person in time, and help each person constitute oneself in time. But there is much additional wisdom in Proust's writing. For instance, because his writing shows that perception is not an external relationship, Proust is able to unconceal truths about Marcel's (and every person's) perception of musical beauty. Such a perception is delicately articulated and illuminated when, in the novel, Morel plays Vintuel's sonata. Another example that Proust eloquently describes is Marcel's immediate perception of the beauty of Elstir's painting of a cliff above the sea. We shall not present Proust's detailed discussions of the beauty of a musical composition or a painting, but we can add that appreciation of beauty is frequently sensual and pre-reflective.

Thus, in his writing, Proust was constantly aware of the independent actions and interactions of a person's mind and body. Marcel's tasting of the madeleine soaked in a decoction of lime flowers—not an action of his intellect nor a judging of certain sense data, simply the taste—brings forth in his body and mind the vivid detailed memory of Combray and its people and surroundings. As mentioned, this example is but one of many found in *Remembrance of Things Past*. We hold that one of Proust's greatest contributions to human thinking is his articulate and detailed descriptions of the memories and meanings that become unconcealed and present to us through our simple bodily engagements and perceptions. By these descriptions, Proust has brought forth from concealment one of the major truths concerning perception and memories of the past—that these acts of consciousness transcend both empiricism and rationalism. It is not surprising that this truth very much accords with the ontology of Merleau-Ponty.

We repeat the major finding of this digression. In *Remembrance of Things Past,* Proust articulately described manners by which a person's body and consciousness endow his or her perceptions with memories and with meaning. These articulate and wise descriptions do not accord with the theories of human perception and thinking presented by the empiricist or by the rationalist. Hence the theories are at fault. Put differently, we hold that Proust's masterpiece describes important truths about perception, and that in Proust's writing, perception retains much of its breadth, depth, and wealth.

Moreover, as in the case of the madeleine that Marcel tasted, Proust is aware of the mystery of human perception, and he partially describes moments in which this mystery emerges. Merleau-Ponty seems to be aware of this mystery. In contrast, the essays of the empiricist or rationalist philosophers which describe perception see no mystery. Hence, they describe perception in a narrow impoverished manner. They totally ignore the wealth of possibilities which arise in a person's mind and body when that person interacts with the world. As Merleau-Ponty puts it, the presentations of both the empiricist and the rationalist "are equally independent of the action of the mind."[4] From Proust and Merleau-Ponty we can learn that they are also equally independent of the life and action of the human body.

Another distortion found in both empiricist and intellectualist thinking is the firm belief in the validity of the subject-object dualism. The subject-object dualism holds that human beings are subjects and the beings in the world are objects. This dualism has been prominent in scientific and philosophical discourse at least since the sixteenth-century scientific revolution and the writings of Descartes. Empiricists, rationalists, and scientists accept subject-object dualism as the natural and the true manner of relating to the world. Merleau-Ponty repeatedly indicates that such dualism is, at best, problematic.

A firm belief in subject-object dualism, Merleau-Ponty adds, currently determines the research of the natural and behavioral sciences. It also determines much philosophical thinking. A major problem with a philosophical approach that accepts the subject-object dualism is that it examines human beings from an external perspective. Consider the rationalist philosophers. Like Descartes, many a rationalist philosopher will begin with an examination of one's own consciousness and cogito. After a detailed examination of the cogito, and on the basis of his or her findings during the examination, the rationalist will present certain arguments which instruct us to embrace an external acceptance of the world and of the many beings in it. During this research, the rationalist will not discuss or examine the pre-reflective consciousness and its manner of perceiving beings that exist in the world.

Such an approach leads to an interesting outcome, which belies the seeming contradiction between rationalism and empiricism. The outcome is that many of the rationalists embrace the external perspective that is central to empiricist writing. They have adopted the empiricist manner of viewing the human entity and the world as separate beings.

Look closely at the details of the thinking of these rationalists. They generally follow Descartes, who, at first, doubts the existence of the world, but cannot doubt his own existence. From his own existence, and from the rationally proven existence of God, Descartes accepts the external world as he encounters it. He recognizes the world as external to him and existing, and as full of independent objects. Such an acceptance of the objects in the world accords with the views of the empiricist. Since many rationalists follow in the footsteps of Descartes, they believe, like many empiricists, that the human entity and the world are separate beings, which somehow can be bridged.

However, as phenomenological and existentialist thinkers have suggested, the rationalist-empiricist separation between the human entity and the world is false. The truth is that we are bodily bound to, involved with, and engaged in the world. Or, as we have already stated, the human entity is a Being-in-the-world, as Heidegger put it. Primordially, Merleau-Ponty shows, the human entity, the human body is not an external viewer of the world. Writing with poetic flair, Merleau-Ponty adds truth and vividness to Heidegger's idea:

Our own body is in the world as the heart is in the organism: it keeps the visible spectacle constantly alive, it breathes life into it and sustains it inwardly, and with it forms a system.[5]

Such a relationship between the human body and the world firmly rejects the prevailing subject-object dualism. We return to this illuminating citation later in this chapter and also in the following chapters.

In the Introduction we mentioned that, according to Merleau-Ponty, the rejection of the subject-object dualism entails the rejection of other accepted dualisms; among them are person-world, body-soul, and nature-culture. He holds that all of these dualisms determine reality in a false manner. Yet, he acknowledges that they have become rooted in much contemporary thinking and action. As he repeatedly states, the truth is that body and soul are firmly and delicately interwoven in a person's being, much as nature and culture are firmly and delicately interwoven in a society's history and life. Hence, we again state: The view of reality that unquestionably accepts firm dualist distinctions is false. It conceals truths about human perception and relations to the world; it also obscures thinking. This false view also creates many problems for any person who wishes to establish an honest and just relationship with the beings and the species with whom we share the world.

For instance, following Descartes, many of the rationalist dualists consider everything that is natural in a human being, everything that is linked to the human body, as secondary to reflection and to the human essence. The human essence, they add, is centered upon a person's ability to think rationally. That essence, the ability to think, some rationalists explain, is the basis of culture; yet, they add immediately, that thinking is based on language which is central to a culture. Put differently, they explain correctly that the ability to think and to act rationally is probably a natural endowment, but, they add, rational thinking can only emerge within a language and a living culture.

Despite the specific truths to which these rationalists point, we assert, but shall not attempt to fully prove, that there are grave dangers in such a confined approach. Rational thinking, as history has repeatedly revealed, does not ensure a worthy human existence. A worthy existence, according to Kierkegaard, Nietzsche, Heidegger, and other thinkers, requires more than rational thinking. It requires caring about other beings, and living in a manner that brings forth at least some of the elements of human grandeur.

We should add that history and literature abound with examples where human beings found rational justifications for committing unrestrained cruel and brutal acts. Think of Dostoyevsky's Raskolnikov in *Crime and Punishment* and his rationally justified murder.[6] Unfortunately, Raskolnikov is a mere example of what occurred in twentieth-century history, and still occurs. Among the political acts that attained rational justification in the twentieth century were ruthless and savage wars, instances of genocide, oppression of entire populations, and exploitation of millions of human beings. These acts were not worthy and did not express the grandeur of humanity.

The many dangers for human existence and society arising from an unskeptical acceptance of rationalist dualism and of rationalist thinking have been repeatedly discussed by twentieth-century thinkers. Albert Camus and Nikolai Berdyaev are two among quite a few such thinkers.[7] However, these and other thinkers rarely related the dangers of rationalism and of subject-object dualism to our everyday relationships with other natural beings and species. Hence, we shall briefly focus on a few of these dangers.

One danger emerges when persons uncritically accept the rationalist dualist distinction between nature and culture. The distinction will allow some of them to conclude that nature is there to be dominated and exploited. Specifically, such persons may embrace the belief that, if nature is secondary to culture and to rational thinking, as long as a person is dedicated to culture and to thinking, there is no harm in an uncurbed exploitation of nature. Some of these armchair rationalists might add that there are no rational grounds for criticizing the rampant destruction of many natural beings, say spiders and salmon, brought about by the actions of human beings, and especially by human actions in the past three centuries. A few of these dedicated rationalists might add that they do not see any rational grounds to reject the exploitation of nature that accords with capitalist doctrines. Others might state that they see no rational reason to reject genetic modification of plants and animals. Put bluntly, such armchair radical rationalists would not find much fault in many of the brutal acts in which human beings destroyed and are continuing to destroy the environment, acts in which human beings terminated the existence of other beings and of species that share the world with us.

To bolster our assertion concerning the dangers for nature that stem from rationalist dualist thinking, we pose the following rhetorical questions. Why should a rational human being—who believes that his or her own body is secondary to his or her thinking essence, who holds that nature is secondary to culture and to one's own well-being within one's culture—be perturbed by the rampant destruction of beings and species in our natural environment? Or, to be specific, why should a person who stresses rationality and culture care if many thousands of undiscovered or supposedly "not needed" species of plants and animals, say, of lichen or beetles, vanish each year from the face of the earth?

We can now ask a key question. Does rational philosophy provide a sufficient philosophical ground that will encourage the thinker to establish and to justify a struggle against the unrestrained destruction of natural species? Our answer to this question is: No.

The reason for our answer is simple. Once rationalist dualism is accepted, there is almost no place for a belief in the significance of sharing one's being and the world with other beings. Put otherwise, the accepting of rationalist assumptions and of the subject-object dualism demands viewing the minds of persons as permanently detached from their body, from nature, and from the world. This conclusion, as Merleau-Ponty repeatedly shows, is wrong.

Rationalism, however, has additional faults; we mention two. First, the rationalist philosophical approach condemns human beings to an unwarranted aloneness; it also accepts, as a fact of life, human irresponsibility for the fate of other beings in the world. Care for other beings, as an ontological aspect of human existence that Heidegger described, is not included in their understanding of human reality. In addition, sharing with the other, as an ontological possibility of human existence—a possibility that Merleau-Ponty shows emerges in perception—does not emerge in the rationalist scheme of the world.

Thus, because of its basic assumptions, rationalism constitutes an impoverished and often superficial thinking. For instance, the rationalist will never be able to comprehend much of the wisdom in Martin Heidegger's profound essay "Letter on Humanism." Note that in "Letter on Humanism," Heidegger not only criticizes past and contemporary thinking on humanism, he also proposes a humanism that will do justice to the grandeur that human beings can strive to attain. Heidegger also wishes to bring forth important truths from concealment concerning human existence and to enhance the goals of humanism.

One such significant truth emerges from concealment when Heidegger states in that essay "Man is the shepherd of Being."[8] When faced with such a statement, which describes the ontology and existence of the human entity and discloses the responsibility facing human beings, the rationalist is struck dumb. Heidegger's ontology, thinking, and statements are beyond the rationalist's furthest purview.

Learning from Husserl and Heidegger, Merleau-Ponty repeatedly indicated that there are other philosophical approaches to comprehending human existence and perception than those proffered by rationalism. These philosophical approaches relate to the mystery of human existence and of our ability to engage in thinking. These approaches also disclose truths about human existence and perception. Rationalism and empiricism will never be able to comprehend these truths or this mystery. In his research, Merleau-Ponty presents an approach to human existence and perception that does not accept the detachment of persons from the world that is promoted by the propounders of the subject-object dualism. But such detachment is central to rationalist philosophy.

To conclude this point, note that in the earlier citation from *Phenomenology of Perception* Merleau-Ponty wrote: "Our own body is in the world as the heart is in the organism." We may add that no human heart willingly makes its own body into an object from which it is detached. Nor does a heart willingly destroy its own body as the human organism is destroying much of the natural world.

Why has Merleau-Ponty's pointing to the distortions of rationalism and empiricism often been ignored? We do not know. We believe that you need only look around you with clear vision, you need only think courageously, and you will, at least partially, witness and comprehend the widespread distortions of reality promoted by rationalism and empiricism—distortions to which Merleau-Ponty repeatedly pointed. When such comprehension occurs, you may also perceive some of the sad outcomes for human thinking and living that come into being when the rationalist or empiricist approaches are broadly adopted. These sad outcomes especially emerge when rationalist and empiricist approaches are adopted without being questioned.

Put differently, seeing the distortions that are found in empiricism and rationalism requires the courage to lucidly perceive mistakes in accepted theories. But instances of such courage are quite rare in the world, in academia, and in the writings of many environmentalists. This book is our attempt to muster the courage to look at those grave distortions concerning perception which were pointed out by Merleau-Ponty. We also have decided to enumerate some of the sad outcomes—especially in our relationship to the beings of nature—to which these distortions lead.

Consider, once again, a few such sad outcomes. A person who adheres to rationalism can wrongly assume, or, at times, his or her language and its scientific and philosophical orientation will encourage the person to assume, that we human beings exist so as to dominate the earth. Hence, there is no need to share the earth with other species or beings. Such assumptions also attempt to conceal the fact that we human beings are sojourners on the face of the earth. In truth, the situation is much worse. Many a rationalist thinker, while using only rational language, and while adamantly ignoring the mystery of the Being of beings, will frequently suggest that we rational beings are the lords of the earth. Consequently, for us rational lords of the earth, the rationalist will conclude, there is little need to encounter or to meet the other beings in the world in any kind of dialogue. A person needs only to know how to deal with these beings in a rational manner, how to manipulate them as resources for one's own benefit. Need we add that the rational belief that you must only learn how to deal with and manipulate the encompassing human beings and beings of nature—and not encounter these human beings and beings of nature, or share the earth with these beings—is central to the reigning capitalist way of life?

How would the rationalist respond to our repeated criticism? One widespread response argues that we human beings are endowed with the capability of rational judgment. Consequently, we can relate rationally to other beings and to species of nature. Furthermore, the rationalist may announce, we can change our supercilious lordly mode of behavior toward nature if we rationally and wisely judge the unwarranted destruction and devastation that human beings bring about in nature. We can also wisely judge, the rationalist will add, the dire state of nature that, say, our unchecked use of lethal chemical compounds in agriculture and in industry has brought about.[9]

We submit that there are elements of truth in this brief response of the rationalist. Yet, in order to perceive some of these elements of truth clearly, it is wise first to consider Merleau-Ponty's rejection of the rationalist concept of judgment. In his rejection of this concept of judgment, he begins with ontology.

Merleau-Ponty shows that for many rationalists, be they psychologists or philosophers, a rational judgment is a way of explaining reality. Descartes, for instance, believed that the judgments of his cogito can solve the major problems of the universe; as mentioned, among these problems Descartes included the proof of his own existence, of the existence of God, and of the existence of the world with its beings. The same is true, with significant variations, of the underlying approach adopted by Spinoza, Leibniz, and many other intellectuals up to Kant, Hegel, and beyond.

This quest for a rational proof and explanation, the rationalist may suggest, arises from the human need to establish order in the chaotic effects that are brought upon us by our sense impressions. Thus, in a roundabout manner, like the empiricist, the rationalist is dividing the workings of the human consciousness to sense impressions and judgments. The sense impressions that we receive from the world are, according to this approach, chaotic; hence a person needs judgment in order to establish what truly is out there in the world. Even if the rationalist agrees that at the level of sense impression there is some judgment, say in our sight or hearing, he or she still retains the division between sense impression and judgment. This division, Merleau-Ponty holds, is erroneous. He adds that, since sense impressions are attributed to my body and judgment is attributed to my soul, the distinction mirrors the accepted stark division between body and soul; it also mirrors the accepted radical division between nature and culture. Both divisions are erroneous.

Rejecting both the rationalist and the empiricist description of judgment, Merleau-Ponty argues that judgment is central to any act of perception during the act. I immediately grasp that the newspaper is upside down and that to read it I have to turn it over. On a foggy day, I suddenly see that the barely perceptible brick-like elongated structure, reaching upward beyond the mist, is a church steeple. I immediately smell the onions frying on the stove, and not an odor that needs defining by judgment. I immediately hear the patter of rain and not a drumming on my window that needs to be judged in order to be determined as rain. In short, perception *is* a judgment. Hence, the stark division between judgment and sense experience, that has been promoted by the rationalist, leads us astray.

Merleau-Ponty agrees that there are moments in which I retreat from perceiving and decide to judge a specific matter that I have encountered. But these are rare moments; they occur when a person attempts to be detached from immediate involvement. They are moments of reflection or of thinking, not of perception. Thus, even the basic act of judgment, which emerges in every instance of perception, is based on the fact that the human entity is primordially involved in a world that he or she shares with others. In contrast, the rationalist's concept of judgment ignored this primordial involvement in the world; it detaches the person from the world that he or she share with others in order that the person may judge. It ignores and rejects the truth that every moment the human entity perceptually and existentially shares the world with other beings.

Again, we point to the significance of Heidegger's locution: Being-in-the-world. This locution discloses the primordial involvement of the human entity with the world and with the beings in the world. One immediate outcome of this primordial involvement in the world—an involvement which exists through the act of perception and other acts, is that the human entity can never detach itself from the world.

A major reason that such detachment is virtually impossible is that each person, as Merleau-Ponty shows, is an integral part of every field of perception that his or her perception opens in the world. (We explain more exactly what is meant by a field of perception in Part Two; here, as before, we rely on the intuition of the reader.) Yet, we repeat: It is precisely this primordial involvement and this Being-in-the-world of the human entity that the empiricist and rationalist blatantly ignore. Both empiricism and rationalism are based on the detaching of what they call the perceiving thinking subject from the object that it perceives; in most cases these objects are beings that exist in the world.

Our emphasis of the error of detaching the perceiving person from the world is significant since a similar detaching is prominent in the writings of many environmentalists. For instance, the theory that rational thought can lead to a striving for, say, "a sustainable environment," which we criticized in Chapter 1, is based on an act of thought which detaches the thinker from the world, and sets him or her up as a lord of the beings that exist in the world. As such a lord, he or she will determine which species are worthy of living and thriving in a "sustainable environment."

To recapitulate, although there are elements of truth in the environmental theories that the rationalist or empiricist suggest, and that are supposedly based on rational judgment, we have shown that they are problematical. Perhaps the greatest problem in these theories is that they are not ontologically rooted in human existence in the world. These rational rootless theories flutter around us, like beautiful butterflies in a tract of burnt forest.

We can now suggest, on the preliminary level, an additional significant contribution of Merleau-Ponty's ontology of perception. This contribution is presented to us as a baksheesh to the truths concerning perception that Merleau-Ponty brings forth from concealment. The contribution is that, on the basis of the ontology of perception, we human beings can think lucidly about living in a world that we share with other beings and species. Moreover, if we have courage, we can see through a cloud of misconceptions that are promoted by greedy and pernicious individuals. Indeed, despite this cloud of misconceptions, we can see that there is much glory and beauty in sharing the world with other beings and species. How do we learn to comprehend this glory and beauty? The answer is not difficult.

Let us accept Merleau-Ponty's thinking that every person bodily shares every field of perception that he or she perceives with other beings that are in the world. We also accept that this perceptual sharing is primordial and is central to and constitutive of every person's being. This assumption leads us to hold that a person will reach fulfillment and live in truth when he or she grasps the primordial ontology of his or her being and lives one's life in harmony with this primordial ontology. Such a harmonious existence requires living as a person who shares the world with other beings.

Strange as it may seem to persons engrossed in rational approaches to every problem, the idea of fulfillment through sharing is not a radical idea. Many of the great spiritual thinkers of humanity showed that the path to a spiritual existence must include sharing the world with other human beings and with the beings of nature. This fulfillment and spirituality, they indicated, can make human existence just and beautiful and may bring glory. A vivid example of such a spiritual leader is Saint Francis of Assisi. Another example is Socrates, as he is described by Plato.

Note another point that Saint Francis and other spiritual thinkers showed their adherents: A lived sharing of the world should lead each person to a modesty that emerges when he or she relates to the other beings of the world, and especially to living beings. And such a modesty, as many thinkers and poets have suggested, can lead to wisdom. Think of the development of wisdom in the life of King Oedipus in the two plays on this myth by Sophocles; or think of Socrates continually announcing that he knows that he does not know.

We can present the gift of the baksheesh a bit differently. We receive a baksheesh from Merleau-Ponty's thinking, because he repeatedly shows that we human beings, as perceiving bodily beings, do not dominate the world. Rather we meet the beings in the world in a bodily encounter and engagement; this bodily meeting constitutes our everyday existence. Moreover, each human being discovers oneself as a unique person through his or her decisions during one's daily bodily encounters and meetings with the beings in the world. Hence, from a primordial perspective, that is, from the perspective of our initial bodily consciousness of our selves and of the world, we are sharers of the world with other beings, who exist in the world. We human beings daily establish ourselves in a world that we share with other beings, and each person creates his or her particular being in a world established by one's fields of perception. Thus the world has many beings that participate in creating each person's being.

The baksheesh that Merleau-Ponty's writings proffer allows us to state a conclusion that goes beyond his presentation in *Phenomenology of Perception*. The conclusion is that when a person recognizes that he or she daily creates one's personal being in a shared world, he or she has taken an important step toward personal fulfillment. Such a recognizing of one's existential situation may also lead a person to struggle for a more spiritual existence. Learning from Plato and other great thinkers, we hold that such a spiritual existence may include, among other worthy things, struggling for justice, seeking beauty, engaging in love and friendship, the pursuit of wisdom. Merleau-Ponty's baksheesh is a gift that can help persons endeavor to reach the heights of human existence. We return to this baksheesh from a different perspective in Chapter 8.

Thus, in addition to being true, Merleau-Ponty's critique of the distortions of rationalism teaches us that a person can recognize, on the basis of ontology, that he or she can live with integrity in a shared world. Such a recognition may open up a worthy realm of existence for that person. Within this opened realm, elements of human glory and wisdom, in addition to natural grandeur, can come into being. Moreover, persons who enter this realm, as sharers, will often grasp that it is wise to relate with modesty in their relations with other beings, and with the beings of other natural species. Think of Hemingway's Santiago as presented in Chapter 1 and in *The Old Man and the Sea*.

It is unfortunate that many prominent environmentalists do not realize that the rationalist philosophical approach is divorced from an ontology of sharing. These environmentalists may realize that many of the empiricists and rationalists believe that they, as human beings, dominate the world. They will comprehend that the princes of corporate capitalism, and their academic spokespersons, are hardly concerned with sharing the beauty and richness of the world with others. They may also grasp that these princes and academicians are not at all concerned with relating to the environment in a manner that, to borrow a locution from Heidegger, will let the beings in the world be. Yet these environmentalists seem to be fixated on the tenets of empiricism, capitalism, and rationalism. Consequently, as already mentioned, these environmentalists use empiricist, rationalist, and capitalist terms and argumentation as a basis of their views.

For instance, these rationalist environmentalists may argue that there is much monetary benefit in sharing the earth with other species, and that letting the beings on the earth fulfill their being may bring profits to those involved in such schemes. Thus, they will applaud every instance of environmental tourism. However, they will never suggest the existence of an ontology of sharing. They will not intimate that attempting to live in accordance with an ontology of sharing may help wean human beings from a faulty approach to the world and to human existence in the world. Most of them will never reject the approach which considers the earth and the beings in the world as a source and not much more—a source which we human beings can exploit for our own means and ends.

Someone may still say: You are much too abstract. Perhaps you should present an example. What precisely are the mistakes of those dedicated people who accept the Cartesian rationalist approach and the capitalist regime, and still want to save the environment from destruction?

Before presenting two examples, we want to clearly state that we agree that there are many rationalists and supporters of capitalism who are committed to saving the environment from ruthless human destruction. We also agree that many of these rationalist and dedicated environmentalists act courageously so as to halt the deterioration of the environment, and that, frequently, they proffer noble rhetoric. Some of these environmentalists reveal, in their deeds and writings, the passion that is needed in order to fulfill, at least partially, their worthy intentions. Yet we have also found that, however noble their rhetoric, however fierce their struggle, the philosophical premises of these environmentalists are quite banal and faulty. Their thinking lacks a valid ontological foundation. Furthermore, because they never grasped the underlying ontological problem to which we have pointed, the language adopted by these dedicated environmentalists does not respond in depth to the problem of relating to the beings in the world. Here are two examples.

We have already stated that one of the leading and famous spokespersons for preserving the diversity of life on the earth is Edward O. Wilson. As mentioned, Wilson teaches at that so-called fortress of excellence, Harvard University, and is famous in Western society for his dedicated struggle against the ruining of the natural environment. Consider a short citation from his book on biodiversity, *The Diversity of Life*. In this book the reader soon learns that Wilson passionately wants to save the biodiversity of the world from destruction by the results of uncritical human exploitation, and by the rampant destruction that results from ignorance. The citation points to his language and thinking.

Merely the attempt to solve the biodiversity crisis offers great benefits never before enjoyed, for to save species is to study them closely, and to learn them well is to exploit their characteristics in novel ways.[10]

Among other things, we learn from this short citation that biodiversity is a crisis to solve, not something that we should, say, comprehend with wonder, respect, and awe. We learn that the attempt to solve the crisis offers great benefits, which for Wilson are mainly monetary and scientific. We also learn that, if we save a species, it is in order to exploit its characteristics in novel ways. You do not need to be an astute scholar to discern that such an approach is that of the lord who dominates the world, very much like a dictator. We repeat: In the citation, there is no awe or respect for the members of the biological species as beings in themselves. These beings are regarded to be mere means or resources for human ends.

Put differently, there is no hint in this citation from Wilson that, ontologically, we human beings are constituted so as to share the world with other species. There is no hint that this manner of being, that is, our sharing of the world with other beings includes grave and noble responsibilities. One such responsibility is that, to the best of our abilities, we must struggle to let these beings exist as they are, with their uniqueness, without linking their existence to any benefits for us. No such responsibility for other species is mentioned in the citation. In Wilson's book, perhaps one can find fleeting hints as to such a noble responsibility. But he never directly states our responsibility to let the beings of the world be.

In parentheses, we want to add what should be an evident point. An approach that promotes sharing the earth with other beings does not reject our responsibility for human health. For instance, it does not reject our responsibility to eradicate the microbes, viruses, and other pests that cause malaria, tuberculosis, AIDS, or other terrible diseases.

The reader of the above citation and of Wilson's book soon discovers that the vocabulary and the idiom that Wilson adopts is taken from the language of corporate capitalism and its intellectual supporters. Put differently, in order to preach his beliefs about the benefits of biodiversity, Wilson uses this language of exploitation and domination—in the above citation and throughout his book. The results are ludicrous. Any corporate capitalist can insert for the word "biodiversity," in the citation from Wilson, two words, say: "mining diamonds." The capitalist will feel very comfortable with Wilson's sentence after this insertion. It is certainly true, the capitalist will hold, that overcoming the current diamond crisis will bring great benefits to humankind. Furthermore, he or she will add, if we study diamonds (instead of a natural species) closely, and learn them well, we can exploit their characteristics in novel ways.

We believe that Wilson's descriptions about the extinction of natural species, and his statements on the need to save the biodiversity that exists on our planet, are important. Unfortunately, because of Wilson's shallow and myopic ontological stance, because of the corporate capitalist language that he uses, the writings constitute a fig leaf for the greed and lust for power which are currently central to the ruling corporate capitalist regime. Wilson only vaguely hints—if at all—that this greed and lust for power, taken together, determine much of our existence and destroy the thousands of species that are dear to his heart. His assumptions ignore the earlier critique of rationalism; in many respects, they ignore all philosophical thinking.

In contrast to the books published by Wilson, consider the writings of Vandana Shiva, who also has very little philosophical background. Shiva is a physicist by training. Yet in the past two decades, she has become one of the well-known spokespersons against the ruining of the environment by corporate capitalists. Indeed, her writings express great sensitivity to the sad facts concerning the evil onslaught of corporate capitalists on the natural environment—and also the manners by which capitalists disregard the lives of simple people. These writings show that when we are at the mercy of this evil onslaught, we learn not to share our being with other beings on the face of the earth. Consider two paragraphs from one of her essays.

We are in the midst of war, a war unleashed by the global market on the life of people. This war is an everyday war—a war carried out in the name of economic competition, a war of corporations against the people. It is an organized theft of the survival resources of the poorest in order to increase the profits of the most powerful corporations.[11]

Intolerance of diversity is the biggest threat to peace in our times. The cultivation of diversity is in my view the most significant contribution that can be made to peace—peace with nature and peace between different peoples. I say "cultivation" because it has to be a conscious and creative act, intellectually and in practice. It demands more than mere tolerance of diversity, because tolerance is not enough to contain the wars unleashed by intolerance of diversity and difference.[12]

The citations show that Vandana Shiva is much closer to describing the truths concerning the daily threat against biodiversity in the world than Edward O. Wilson. She points out who is today ruining the biodiversity that exists in the world—the powerful capitalist corporations and their many political and academic supporters. She calls the corporations' daily activities a war against the people, in which nature and the people are exploited, oppressed, and at times, destroyed. She explains that this devastation of the environment and exploitation of human beings is done to enlarge the profits of a minority of affluent people.

The above citations imply that these capitalists who ruin the environment, these champions of globalization and free trade, are greedy and evil. They are destroying fellow human beings and our natural environment. Thus, Shiva sees and describes the destruction of the earth as the result of the domination of an evil economical and political regime which emphasizes and extols greed. But, she speaks the language of rationalism.

We wholeheartedly agree with Vandana Shiva's harsh critique of the destruction and devastation wreaked upon the earth and millions of its people by the princes of corporate capitalism and their many supporters. You need only open a newspaper to discover that these princes are supported by many political lackeys and academic sycophants. We agree that Western governments support this capitalist devastation of large regions of the earth, and ignore the ruin of people and of the beings of nature that result from this devastation.

Unfortunately, however, in the above citations and in her other writings, Shiva does not present a philosophical basis for the worthy ideas that she suggests about the human responsibility to share the earth with other persons and species. She provides no ontology of human existence that can counter the dominating and domineering rationalism and empiricism dear to the princes of corporate capitalism. Shiva does not notice that the philosophical toadies of the princes of capitalism use a watered-down rationalism, at times blended with empiricism, in order to justify the evil deeds that she forcefully condemns. Nor does she indicate that an ontology exists which may lead to a more honest and authentic way of existing in the world.

Consequently, we believe that Shiva's writings are more of a rebellion than a revolution. She rebels against the evil deeds of earth's true enemies and the enemies of a majority of humanity—the corporate capitalists and their cohorts of supporters. But Shiva says very little about what we can do to revolutionize our philosophical and everyday thinking about our place on earth and our responsibilities for the earth and its many natural beings.

To recapitulate, this chapter describes some of the ontological distortions that empiricist and especially rationalist philosophies bring about. It suggests that these ontological distortions influence our grasping of our role and responsibilities as human beings in the world. We have indicated briefly that Merleau-Ponty's rejection of empiricism and rationalism and his thinking on perception shows a worthy path beyond these distortions. We believe that in order to proceed along that path, something close to a revolution in our thinking, a revolution that discards the distortions of both empiricism and rationalism, must come into being. We believe that such a revolution in thinking will have many yet unseen manifestations. Hints as to some of these manifestations appear in the following chapters.

In concluding this chapter, we can say that, for a revolution in our thinking to come into being, Vandana Shiva, and other worthy strugglers against the destruction of the earth, must include a critique of the rationalism and the empiricism that the capitalists and their many philosophical justifiers embrace. These strugglers for justice and for the environment must grasp that both rationalism and empiricism are used by the adherents of corporate capitalism in order to promote and justify the greed and lust for power that underlies the unbridled reign of the corporations.

Yes, Shiva and her fellow strugglers must firmly grasp that empiricism and rationalism are frequently used to conceal the widespread greed and ruin that accompany the march toward globalization. They must firmly reject the rationalism that currently governs much of our existence. They must reject both rationalism and empiricism because such thinking is wrong, and because it serves those people who have unleashed the war of the global market against the people, and against the beings that populate the earth.

The ontology of Merleau-Ponty, we have shown, provides a basis for a radical critique of rationalist and empiricist thinking. Merleau-Ponty's thinking repeatedly indicates that our being is founded and develops through engagements with other beings in the world and by sharing the earth with these beings. Hence, his thinking, we believe, may help to launch such a significant revolution.

NOTES

1. See Maurice Merleau-Ponty, *Phenomenology of Perception*, trans. Colin Smith (London: Routledge & Kegan Paul, 1962), 26.
2. Marcel Proust, *Remembrance of Things Past*, trans. F. Scott Moncrieff (New York: Random House, 1934).
3. Ibid., 36.
4. *Phenomenology of Perception*, 28.
5. Ibid., 203.
6. Fyodor Dostoyevsky, *Crime and Punishment*, trans. Richard Pevear and Larissa Volokhonsky (New York: Vintage Books, 1993).
7. See, for instance, Albert Camus, *The Rebel*, trans. Anthony Bower (New York: Vintage, 1991); Nikolai Berdyaev, *Slavery and Freedom*, trans. R. M. French (New York: Charles Scribner's Sons, 1944).

8. Martin Heidegger, "Letter on Humanism," in *Martin Heidegger: Basic Writings* (New York: Harper & Row, 1977), 210. Farrell Krell (ed),

9. One of many studies that describes the destruction brought about by the use of chemical compounds in agriculture is Sandra Steingraber, *Living Downstream: An Ecologist Looks at Cancer and the Environment* (London, 1999).

10. Edward O. Wilson, *The Diversity of Life* (New York: W. Norton, 1992), 306.

11. Vandana Shiva, "Diversity and Democracy: Resisting the Global Economy." *Global Dialogue*, Vol. 1. No. 1 (Summer 1999): 19.

12. Ibid., 20.

3

Insights and Problems in Gestalt Psychology

The findings of Gestalt psychologists have contributed substantially to contemporary psychology. Some thinkers have suggested that the descriptions of Gestalt psychologists have revolutionized our understanding of human perception and behavior. In the past few decades, some of the adherents of Gestalt psychology have developed methods of psychotherapy which, they explain, are based on the prominence of the Gestalt in human interaction. We shall not relate to the Gestalt methods of psychotherapy, which are hardly linked to the theme of this book.

During the twentieth century, many of the findings of Gestalt psychology concerning perception were lauded by philosophers who learned from the phenomenology of Husserl, among them Merleau-Ponty. These findings have significance for the theme of our book because they help to dismantle the false constructions of reality that were developed by thinkers in the empiricist and rationalist traditions. This dismantling of broadly accepted constructions of reality opens the possibility for accepting the ontology of sharing that emerges in *Phenomenology of Perception*. Yet, a word of caution is necessary. While praising the findings of Gestalt psychologists concerning perception, Merleau-Ponty also criticized some of the conclusions concerning human existence and perception that these psychologists suggested.

The term "Gestalt-quality" was coined by the Austrian philosopher and psychologist Christian Freiherr Von Ehrenfels (1859–1932). He first used the term "Gestalt-quality" in a paper that was published in 1890, which is recognized as inaugurating Gestalt psychology. In his paper, Ehrenfels pointed out that, almost always, when a person listens to a musical composition—say, when someone listens to a good orchestra performing Mozart's 40th symphony—the listener hears a melody. We repeat this seemingly evident point. The listener to Mozart's 40th symphony does not merely hear a sequence of musical notes or a sequence of sounds—he or she hears a melody. This melody is a form, a Gestalt, that comes into being when the sequence of notes is played by an orchestra. To be exact, the melody is a formed whole that exists in time. Hence, a person may perceive a melody when he or she hears the sequence of notes that, in the case of Mozart's 40th symphony, the orchestra is presenting. Ehrenfels noted correctly that, for the listener, the melody does not arise from any special stimulus that is added to the sequence of notes. Rather, the melody is there in the sequence of notes that the orchestra is playing and that a person hears.

Someone may ask: What happens if a person tries to break down the melody to which he or she is listening, and to direct oneself to merely hearing it as a sequence of notes? Will he or she hear the melody? The answer, of course, is: No. Hearing a sequence of notes differs substantially from hearing a melody.

To illustrate the difference between hearing a sequence of notes and hearing a melody, we believe that we need only consider a Zen fable. Once there was a student of Zen who decided to catch the flow of a small stream of water in buckets. At the end of the day, the student of Zen had gathered all the water that had flowed in the stream into his buckets. But all he had were the buckets of water; the flowing stream had vanished. The same will be true for a listener who wishes to perceive and identify all the notes in the sequence of a certain melody. If the listener to a melody endeavors to hear each of its individual notes as it appears in the sequence, the melody vanishes.

Gestalt psychologists have shown that almost all of our daily perceptions resemble listening to a melody. In these perceptions, we immediately grasp a Gestalt, a whole, a total form, or a structure. We grasp a cube even if we do not perceive, and can never at the same time perceive, its six equal sides. We do not perceive individual items that constitute a Gestalt—the notes of a melody or the sides of a cube—but rather the whole being or totality. Furthermore, the form, the Gestalt, which we grasp in its totality is not inferred from components or items, much as the melody is not inferred from the notes or the cube from its equal sides. Rather, we perceive the Gestalt as a totality into which the components or items blend. Of course, when we perceive a Gestalt, we recognize that there are specific items that constitute the totality that we perceive, just as when listening to Mozart's 40th symphony, we recognize that the melody is *in* the sequence of notes that we hear.

Gestalt psychologists have revealed that an interesting fact about a Gestalt is that it gives form to a multiplicity of possible perceptions and items. These perceptions do not have to be clear, distinct, and defined for the Gestalt to be perceived. They can be vague, partially obscured, or incomplete. I can perceive a circle, even if it is vague or not complete. I perceive a cube even if I do not see all of its equal sides. I can perceive the steeple of a church through the fog that distorts its shape.

Someone may ask: Is every abstract concept a Gestalt? The answer is: No. In order to be a Gestalt, an abstract concept must give a meaningful form to daily perceptions. Put differently, a Gestalt must relate to concrete events in a person's life.

A new Gestalt can arise suddenly, say, when a person coins a concept that did not formerly exist. For instance, consider once again Heidegger's concept of the human entity as Being-in-the-world. This concept suggests a new way of comprehending the human entity in its everyday concrete interactions. Dasein, or the human entity as Heidegger calls it, is viewed as an existing totality within the world and firmly linked to the world. The concept also indicates the human entity's primordial relationship of engagement to other beings that exist in the world. Thus, Heidegger's concept of Being-in-the-world is a way of comprehending human existence that accords with our perceptions of the human entity as constantly engaged in the world. This new Gestalt has philosophical implications. In previous chapters, we have already stated some of these implications. We held that the concept Being-in-the-world reveals the totality of human existence. This human totality is in stark contrast to the manner that the subject-object dualism, that is central to both rationalism and empiricism, presents the human person. Hence, Heidegger's concept presents a new Gestalt.

The emergence of a new Gestalt may also occur with the coining and the use of concepts that describe the rampant destruction of the natural environment. Consider the title of Rachel Carson's famous book, *Silent Spring*.[1]

The name of Carson's book, we believe, coins a meaningful concept and also presents a new Gestalt. The Gestalt that Carson's concept, "silent spring," suggests is that of silence that we encounter in the springtime in the natural world. Needless to say, this Gestalt of silence stands in stark contrast with the Gestalt of a melody.

What is the Gestalt of a silent spring? What is the totality of silence that Carson describes? It is the silence of a spring without songbirds, without the buzzing of insects, without crickets chirping as the sun sets, without the cries of jays, without squirrels emerging from hibernation and busily seeking their food, and without countless other sounds that characterize the renewal and rejuvenation of natural life that occur in the spring in temperate zones. This Gestalt is meaningful because spring, at least in the temperate regions, has been characterized for centuries as a period during which persons hear the renewed sounds of life of the beings of nature as they interact with each other. Hence, a silent spring is an essential change in our Gestalt of what spring brings forth.

Carson may have chosen this title for her book because a spring of silence, without songbirds, without humming insects, without crickets chirping, suddenly came into being in many areas of the United States that had been widely sprayed with insecticides and other harmful chemicals. She hints, but does not directly state, that the reasons for this silent spring were linked to irresponsible and nefarious decisions and deeds by a specific group of human beings. The nefarious decisions were to conduct widespread irresponsible spraying of chemical poisons upon huge tracts of land and forest—supposedly to kill the Japanese beetle and other insect-pests. But, as could have been foreseen, these chemical poisons eradicated almost all insect life in the sprayed area. In addition, all the songbirds in the sprayed area perished, almost all of the squirrels died of poisoning, and many other animals that populated the area were killed.

Although we assume that most of Carson's readers did not live in these areas where natural life was devastated, and did not perceive the silent spring that she described, the Gestalt of a silent spring seems to have brought to many of them a shudder of rage. Even today, four decades after its publication, *Silent Spring* still arouses anger in some readers when they learn of the killing of countless beings of nature that it meticulously describes. Today, it is recognized that the Gestalt of a silent spring, together with Carson's enlightening writing, helped to shatter the complacency that prevailed among many persons who previously were indifferent to the widespread environmental destruction across the United States. As Carson shows, this wholesale destruction of natural beings and species was performed by chemical companies and supported by the United States government.

After Rachel Carson published her book, suddenly, many people seem to have asked themselves perturbing questions: Why am I doomed to endure a silent spring, in which all the songbirds have vanished? Who brought about this silent spring in which squirrels perish of poisoning and disappear, in which there is no buzzing of wild bees or chirping of crickets, in which there is no hooting of owls? Is there no way to halt the evil people and greedy leaders of chemical companies who are causing these silent springs? As this book indicates, halting these evils is a long struggle, which we personally support. In our writing, we are attempting to unconceal the philosophical foundations that we believe should be the ground for such a struggle.

Some readers may have been nurturing a methodological question: How do you define a Gestalt?

In his books, *The Structure of Behavior* and *Phenomenology of Perception,* Merleau-Ponty often relied upon and discussed findings that were presented by Gestalt psychologists.² Here and there in these books he noted that the Gestalt is a basic pattern that emerges in our sensory field, a pattern which we perceive as a totality and on a background. He provided these statements while mentioning some of the important contributions of Gestalt psychology to our comprehension of the reality of human perception and behavior.³ He also described significant problems that emerged in the thinking of Gestalt psychologists, to which we return. But, in Merleau-Ponty's writings, we did not find a concise definition of the Gestalt.

We believe that an apt and concise definition of Gestalt was formulated by Aron Gurwitsch, who wrote extensively about the links between Husserl's phenomenology and Gestalt psychology. He also referred to Merleau-Ponty, who, in turn, was acquainted with Gurwitsch's writings, and cites these writings twice in *Phenomenology of Perception*. Gurwitsch, who was well acquainted with the history of Gestalt psychology and with Husserl's thinking, suggested the following definition for Gestalt.

It [the Gestalt] is an ensemble of items which mutually support and determine one another. Thus they realize a total structure which governs them and assigns to each of them (as a part of the whole) a function or role to be performed as well as a determinate place in that whole. Each detail exists only at the place at which it plays the role assigned to it by the whole of which it is a part.⁴

This definition fits the examples of Gestalt that we have mentioned: a melody; Heidegger's viewing the human entity as Being-in-the-world; Carson's concept of a silent spring; the cube. The definition also fits many other beings and totalities that we encounter in our daily engagements, from a table set for the Sabbath, to a timetable of departures flashed on a screen in the airport. With this definition in mind, we shall examine some of the insights and the problems of Gestalt psychology that Merleau-Ponty describes and discusses, and their relation to sharing the environment with other species.

Quite early in his critique of the empiricist theory of perception, Merleau-Ponty states that Gestalt psychology goes beyond empiricism and shows the weakness of its foundation. One of the most significant conclusions arising from the detailed studies of perception conducted by Gestalt psychologists is that the constancy hypothesis, which is central to empiricist thinking, is false. We explained in Chapter 1 that the constancy hypothesis holds that there exists "a point-by-point correspondence and constant connection between the stimulus and the elementary perception."⁵ To give just one example, if the constancy hypothesis were valid, when listening to Mozart's 40th symphony, I would always hear a sequence of notes—never a melody.

As stated in Chapter 1, many empiricists held and continue to believe that the constancy hypothesis should be basic to our understanding of perception. We also stated there that this belief is wrong. We can now add that you need only look at what is happening when a person perceives, and you will conclude that human perception immediately grasps a Gestalt, a totality, much as I immediately hear a melody and not a sequence of notes, and much as I grasp that the table is set for the Sabbath—I do not infer this totality from the fact that the tablecloth is white, and that you can discern two candlesticks and a bottle of wine on it. And my grasping of the Gestalt, like my hearing of the melody, or like my seeing of the cube or the steeple of a church tower in the fog, cannot be elucidated with the help of the constancy hypothesis.

Unfortunately, many empiricists and scientists continue to believe that the constancy hypothesis must and should serve as the basis for understanding and researching human perception. An especially disturbing fact is that many twentieth-century empiricists and scientists, who had access to the findings of Gestalt psychology, ignored the important insights of Ehrenfels and other Gestalt psychologists who followed in his footsteps.

Put differently, even today many thinkers and scientists disregard the valid implications concerning perception that arise from the findings of Gestalt psychology. For instance, they ignore the implications arising from the fact that in our elementary perception of a symphony or a folk song or a popular hit, almost always we do not hear notes, but rather a melody. What is more, we perceive the melody, the Gestalt, even if the orchestra skips a note, much as we see a cube even if we do not perceive its six equal sides. These findings immediately indicate that the constancy hypothesis is untenable.

Yet, since many scientists and empiricists refuse to abandon the constancy hypothesis, we are faced with a weird situation. Although the findings of Gestalt psychologists seem to have given the constancy hypothesis its death blow, in scientific and philosophical discourse it is still alive and thriving. Today, the constancy hypothesis still serves as an accepted foundation for much empiricist thinking and much scientific research.

Why is this major finding of Gestalt psychology, which accords with Husserl's phenomenology, ignored? Why is the constancy hypothesis still alive and thriving?

The answer to these questions is far from being distinct or simple. What is more, although Merleau-Ponty indicated how Gestalt psychology discloses the poverty of empiricism, he did not relate directly to these two questions, probably since they lead beyond his research on perception. We shall respond briefly and partially to the questions, because the answer has relevance for the theme of our study.

We shall suggest two incomplete answers. We shall not endeavor to prove these incomplete answers, since any such attempt to enlarge on the answers would take us too far astray. The first answer is that scientists have stated that the constancy hypothesis does help them in their research.

Specifically, the constancy hypothesis may provide the foundation needed in order to discover specific limited truths concerning the workings of the senses and other parts of the human body. It probably helps many scientists who are working to develop specific medicines. What is, perhaps, most important is that scientists *believe* that the constancy hypothesis has helped them uncover truths concerning the interactions of human senses with the world. Hence, they have placed the constancy hypothesis at the basis of some of the scientific paradigms that deal with the senses and with perception.

In this Garden of Eden, however, there is a serpent. Look closely. We agree that the findings of many scientific paradigms, say, the paradigm that deals with reflexes, are partially based on the constancy hypothesis. But these findings are all external to the fact of human consciousness and to its workings. For instance, they can contribute nothing to our understanding of intentionality. In brief, the scientific paradigms can say nothing of significance about human consciousness or about how a person perceives a melody, or a cube, or a silent spring, or a steeple in the fog.

Unfortunately, the fact that an external scientific viewpoint cannot teach us anything about human consciousness does not bother the scientists who rely on the constancy hypothesis. They refuse to transcend their scientific paradigms and the constancy hypothesis. They do not perceive the blessing that may come together with their eating from the apple of knowledge that the serpent offers—an apple that will show them truths concerning the limitations of external findings and the errors found in the constancy hypothesis. From this blessing, elements of which can be found in the *Phenomenology of Perception,* they may obtain, at least, additional knowledge about the workings of human consciousness.

The second answer to the above questions has been briefly mentioned in previous chapters. We believe that the subject-object dualism, a dualism that the constancy hypothesis supports, lends legitimacy to the non-sharing materialistic political and economic regimes that prevailed in the twentieth century. Unfortunately, today, in the first years of the twenty-first century, these non-sharing approaches continue to prevail and determine human existence and the existence of other beings in the world. Yes, today corporate capitalism reigns almost unfettered. These twentieth-century regimes, led by the princes of corporate capitalism and their many academic adherents, among them so-called philosophers, were not supportive of any thinking or any findings that challenged the subject-object dualism. Hence these princes and academicians refused to see that empiricism is an impoverished understanding of human interaction in reality, and that the constancy hypothesis is false.

It is therefore no wonder that the materialist-oriented princes of capitalism that reigned in the twentieth century, and many of their academic lackeys, were indifferent to the findings of Gestalt psychology concerning perception. It is also clear that responding indifferently to the truths that were unconcealed by Husserl, and the existentialist philosophers who learned from him, served their approach to human existence on the earth. We may be giving undue intellectual credit to these capitalists and their adherents, but they seem to have intuited something about the Gestalt findings, and the existential truths that Husserl and other philosophers unconcealed—they seem to have sensed that these findings and truths challenged, rejected, and rebuked the narrow non-sharing Weltanschauung that they embraced. Note that these non-sharing materialistic regimes dominated human existence during the period that the findings of Gestalt psychologists revealed that the constancy hypothesis is wrong.

Merleau-Ponty was among the thinkers who were ignored by the princes and pundits of capitalist materialist approaches in the twentieth century. Again and again he announced that the findings of Gestalt psychology unconceal worthy truths about human perception and existence. He often showed that these truths repeatedly disclose the poverty of empiricism and of rationalism. His examples that criticize empiricism and reject the constancy hypothesis are often taken from the findings of Gestalt psychologists. Yet, for many scientists, empiricists, and rationalists, Merleau-Ponty's writings were like a stone that is thrown into a pond but creates no waves.

Concerning the Gestalt psychologists, there is an additional problem. In *Phenomenology of Perception,* Merleau-Ponty's major goal is to present a clear and enlightening phenomenology that focuses on perception and its links to human existence. He soon discovered that responding to this challenge requires that he reject some of the ideas proffered by Gestalt psychologists. He primarily disagreed with those suggestions that Gestalt psychologists proposed in order to describe the foundations that underlie their findings.

Merleau-Ponty's basic criticism of the Gestalt psychologists is that they are not aware of the philosophical implications of their method and findings. Gestalt psychologists seem to have read nothing of contemporary continental philosophy. Hence, they did not grasp that their findings challenged empiricism and rationalism. Nor did they comprehend the philosophical implications that arise from Husserl's phenomenology and from their findings. We have already mentioned Aron Gurwitsch as one of the philosophers who learned much from Husserl; we also stated that Gurwitsch was among those who clearly showed that many of the findings of Gestalt psychologists are in accord with Husserl's phenomenology. However, Gurwitsch also described the fact that Gestalt psychologists learned nothing from Husserl's phenomenology.

The ignoring of philosophy by Gestalt psychologists had grave implications for the foundations that they proposed for their theory. In the first chapters of *Phenomenology of Perception,* Merleau-Ponty accused the Gestalt psychologists of attempting to describe their findings within the context and the assumptions of accepted scientific realism. He argues that the propounders of the Gestalt theory accepted the determinate universe of natural science as given, and as a foundation for their findings concerning human perception. On the basis of this given, they endeavored to fit Gestalt psychology into a scientific and philosophical Procrustean bed that excludes many facts concerning how human consciousness perceives beings. We have already shown that these facts about human perception, which rely on findings of Gestalt psychologists, reject the constancy hypothesis and with it many of the accepted scientific explanations of perception. Indeed, the facts and findings presented by Gestalt psychologists point to basic problems in the accepted scientific way of viewing human existence and perception.

But, lo and behold: Merleau-Ponty was right. As a group, Gestalt psychologists refused to comprehend the possibility that the implications of their findings reject the accepted scientific explanations.

Consider our perception of an object at a distance, or what Merleau-Ponty described as our perception of depth. While looking at the facts, Merleau-Ponty applauds the findings of Gestalt researchers. Listen to an example of his laudatory evaluation of sections of Gestalt theory.

Gestalt theory has clearly shown that the alleged signs of distance—the apparent size of the object, the number of objects interposed between it and us, the disparity of retinal images, the degree of adjustment and convergence—are expressly known only in an analytic or reflective perception which turns away from the object to its mode of presentation, and that we do not go through these stages in knowing distances.[6]

Despite his remarks praising their findings, Merleau-Ponty accuses Gestalt psychologists of embracing superficiality and betraying the truths that they themselves unconcealed. He argues that they immediately repudiate the implications of their findings. For instance, Gestalt psychologists seem to be uncomfortable with the supposedly non-scientific implications that may arise from their findings concerning our perception of distance or depth.

Consider one example of such a betrayal, presented by Merleau-Ponty. Gestalt psychologists hold that the alleged signs or reasons of the perception of distance that are enumerated in the above citation are not signs or reasons, as their research has discovered. The question arises: What are they? The answer provided by Gestalt psychologists is that the alleged signs and reasons are the *causes* of this perception of distance. But what do they mean by causes? The Gestalt psychologists do not elucidate. Hence these so-called causes are meaningless.

Furthermore, Merleau-Ponty argues, with such a statement the Gestalt psychologists have returned to an explanatory psychology and to an unwarranted naturalism. They have abandoned the descriptive psychology which is their birthright and the source of their enlightening discoveries concerning human existence. Ehrenfels described the fact that we hear a melody. Period. He did not seek causes. Thus, the Gestalt psychologists seem to have forgotten that it is precisely their descriptive psychology which allows them to unconceal truths—truths that come into being when they teach us to look at the totality, at the Gestalt. Merleau-Ponty straightforwardly and severely condemns the fact that Gestalt theorists return to explanatory naturalistic approaches: "by this very fact it [Gestalt psychology] betrays its own descriptions."[7]

Later in his *Phenomenology of Perception,* in his discussion of our perception of depth, Merleau-Ponty returns to this criticism of Gestalt theory. We shall not follow all of his arguments.[8] The important point that Gestalt psychologists overlook, he argues, is that both depth and distance are *not* implied from specific signs. Nor are they a result of certain cerebral functions. Distance and depth, Merleau-Ponty explains, are perceived directly by a person who is in a specific situation in which the perception of depth or distance are relevant to his or her perception. Distance and depth emerge in certain fields of perception.

An instructive summary of both Merleau-Ponty's and Gurwitsch's criticism of Gestalt theory was presented by M. C. Dillon, who writes:

The point of that critique [by Merleau-Ponty and Gurwitsch] is that natural entities in the objective world (including human physiology and its processes) cannot serve as the basis for the phenomena of consciousness because it is through the latter that the former are constituted and known. In short, since the objective world is constructed on the basis of the phenomenal world, it is a mistake to reverse the order and try to explain the process of constitution in terms of the objects constituted.[9]

At this point, someone may ask: Why is Merleau-Ponty's learning from and criticism of Gestalt theory important for the theme of your book? What does his relation to the findings of Gestalt psychologists suggest about our responsibility to share the earth with the beings and the natural species that populate it? Our answers to these questions remain on the theoretical level.

One immediate answer, which we have already mentioned, is that the findings of Gestalt psychologists have revealed some major errors of the broadly accepted subject-object approach to our understanding of reality. Gestalt psychologists have repeatedly shown that we perceive totalities and relate to totalities, that is, we perceive a Gestalt and relate to that Gestalt. These simple findings reject both empiricism and rationalism—two philosophical theories which described perception erroneously. As we have shown in previous chapters, these philosophical theories are based on the subject-object dualism, which does not lead to principles of sharing or to the practice of sharing the world with other beings. Worse. We have already mentioned that the principles of empiricism and rationalism have served as the foundations for philosophical and practical approaches that allow human beings to engage in the relentless plunder of the earth and to wreak environmental devastation in many areas and regions. Hence, showing the weakness of these philosophical approaches is imperative.

However, there is also a worthy contribution to sharing the earth that emerges from Gestalt findings. Our perceiving of a Gestalt means that human beings are engaged in a world and not passive receivers of perceptions. Put succinctly, Gestalt psychology shows that our involvement in the world differs substantially from the passive involvement suggested by the principles of empiricism and rationalism. Gestalt psychologists have shown that, on the level of primary perception, persons give meanings to the totalities that they perceive, and perceive totalities that are meaningful. But perceiving a Gestalt, and giving meaning to a Gestalt at the moment of perception, means again that on the primordial level of existence each human being shares one's being with beings in the world while perceiving. In the case of a Gestalt, by giving meaning to a totality, the perceiving person shares his or her being with the item that is perceived, and with the background of that item.

Another significant answer to the earlier questions is that Gestalt psychology leads naturally to the concept of a field of perception which is central to Merleau-Ponty's phenomenology of perception. We have already mentioned that the perceiving person partakes of the field of perception and bodily shares it with other beings. Part Two of this book describes and discusses the field of perception in detail, and its relevance for sharing the earth with other species.

Thus, with the help of the findings of Gestalt psychologists, again we find that our elementary perception is a manner of sharing the world with the beings that we perceive. By perceiving these beings as totalities, by endowing them with meaning in the act of perception—in short, by perceiving Gestalts—we human beings establish the world in which we daily exist. For this insight, we should thank the Gestalt psychologists. However, as Merleau-Ponty has shown, it is unfortunate that the Gestalt psychologists did not grasp the profound ontological implications of their important discoveries. It is also sad that they did not comprehend that their findings have significant implications beyond psychology.

NOTES

1. Rachel Carson, *Silent Spring* (London: Penguin Classics, 2000).
2. See Maurice Merleau-Ponty, *The Structure of Behavior*, trans. Alden L. Fisher (Pittsburgh: Duquesne University Press, 1983) and *Phenomenology of Perception*, trans. Colin Smith (London: Routledge & Kegan Paul, 1962).

3. For instance, see Merleau-Ponty, *Phenomenology of Perception,* 11.
4. Aron Gurwitsch, *Studies in Phenomenology and Psychology* (Evanston: Northwestern University Press, 1979), 25.
5. See Merleau-Ponty, *Phenomenology of Perception,* 7.
6. Ibid., 47.
7. Ibid.
8. Ibid., 257–60.
9. M. C. Dillon, *Merleau-Ponty's Ontology* (Evanston: Northwestern University Press, 1988), 69–70.

4

The Human Body as a Phenomenological Source

In his discussions of the human body, Merleau-Ponty frequently goes further than the thoughts presented by other thinkers who learned from Husserl's phenomenology. He also differs from most major Western philosophers, such as Augustine, Descartes, Spinoza, and Hegel, who rarely dedicated their thinking to the human body, and did not view the human body as a phenomenological source. In contrast to these great thinkers, Merleau-Ponty studied the human body as central to perception, and hence as central to our existence and our engagements in the world. In these detailed studies, he endeavored to obtain a phenomenological understanding of the human body.

In this chapter, we dwell briefly on some of Merleau-Ponty's insights concerning the human body as a phenomenological source. We especially discuss some of those insights that have relevance to the theme of sharing the earth with other beings.

Hold it! someone may say. Before you begin, please explain. What do you mean by the phrase "the human body as a phenomenological source"?

Our answer is based on our learning from Merleau-Ponty and from other thinkers. By that phrase we mean that the human body is the source of, and is crucial for, many of the primal and lasting ontological relationships that human beings establish with other beings of the world. We have already indicated that many of these ontological and significant human relationships cannot be explained by causality, or by other external explanations. They are phenomenological relationships in which a human consciousness bodily perceives a being or a phenomenon within a field of perception, lives with the perception of that being or phenomenon, and relates to other beings on the basis of his or her perceiving.

Love is one such manner of bodily involvement in which a person relates bodily and with one's consciousness to the beloved. In true love, the human body is a phenomenological source of a person's joyfully perceiving of the beloved and relating to him or her lovingly. The body is also a source for the possibility of comprehending the phenomenon of love. Without the body, human love of a specific human beloved does not exist. Discussing the human body as a phenomenological source of love would take us too far afield. However, we can mention three well-known novels in which the reader will find a detailed and enlightening description of the origin of love in a person's bodily consciousness. The novels we have in mind are Leo Tolstoy's *Anna Karenina*, D. H. Lawrence's *Lady Chatterly's Lover*, and Ernest Hemingway's *For Whom the Bell Tolls*.

Merleau-Ponty explains that space is another such phenomenon whose source is in the human body. He shows in great detail that my consciousness bodily perceives space and establishes space. I bodily live in space on the basis of that establishing perception. We shall return to Merleau-Ponty's discussion of space.

We wish to emphasize that, in his philosophical quest to obtain an understanding of the human body as phenomenologically engaged in the world, Merleau-Ponty was swimming against the stream. A brief overview of the history of Western philosophy reveals that few philosophers have studied the human body, its daily engagements, and its manner of being in the world. Even existentialists such as Kierkegaard and Heidegger did not dedicate much thought to the human body. In addition, few past thinkers have been concerned with the way we bodily grasp a phenomenon, such as space, and relate to it. This ignoring of the body is an interesting phenomenon, because each of these past thinkers had a body and lived a bodily existence. It seems, however, that these eminent thinkers purposely ignored their bodily existence.

While refraining from studying the human body and its manners of phenomenologically perceiving beings and relating to them, many thinkers of the past confined their own body into what Merleau-Ponty called an anonymous and passive being. By this locution he means that many persons, including these thinkers, while diligently pursuing their scholarly endeavors, drag their own body along with them "making itself anonymous and passive."[1] After reading Merleau-Ponty and scanning some of the writings of other thinkers, we agree. We find it astounding that so many eminent thinkers in the past centuries ignored their own body and its centrality to their existence and to their perceiving phenomena and establishing relations in the world. Perhaps this blatant ignoring of their own body, while they discussed illuminating philosophical themes, may partially explain why many such illustrious thinkers have relegated the human body to the role of a rather insignificant biological object or process—in which the fluttering human soul is found.

Consider again the broadly accepted subject-object ontology, which, as mentioned in previous chapters, has roots in modern philosophy that commences with Descartes. According to this thinking, the human body is no more than a living object to which a subject with a mind is attached in some obscure manner. Here and there Descartes hinted that we can compare a living body to a machine. Yet, he added, this specific machine has a mind. Descartes' comparison is still with us. From our scant readings, we suspect that the assumption that we can compare the human body to a machine, in which chemical interactions and processes take place, underlies much medical research that has been performed and currently is being performed on the human body. Thus, within the medical sciences, the influence of Descartes' thinking on the body has hardly faded. Furthermore, for many rational philosophers since Descartes, and for many scientists and common people, the human body is merely a living machine-like object in which is found a subject, that is, a mind or a soul. We have already indicated that Merleau-Ponty has shown that this dualist approach is superficial and wrong.

During the past three millennia, some astute thinkers, including Socrates, have suggested metaphysical or theological explanations and justifications for what they believed to be: the independent existence of the soul or the mind. These thinkers believed that the soul which they discuss resides as a separate independent entity in the human body. The soul is eternal, but it remains embodied as long as the body is alive. We shall not present or evaluate any of these metaphysical and theological explanations and arguments, or their underlying assumptions. However, one result of the endeavors of these thinkers is evident. In the past few centuries, their theological and metaphysical ideas and reasonings concerning the human soul have relied upon, and frequently strengthened, the subject-object ontology. Another result that arises from their emphasis on the soul as a separate entity is that, among many adherents of this approach, it has become acceptable to relate to one's body as a passive anonymous being.

Our survey of environmental studies disclosed an additional problem concerning the human body. Many scientists and laypersons who are concerned with preserving the diversity of nature, and who have written worthy studies on the need for such preservation, have not related their thinking to ontology and to the ontological status of the human body. They have not viewed the body as a phenomenological source of our relationships to the beings of nature that we encounter. They have not attempted to comprehend and to study the continual involvement of the human body with other beings in the world, an involvement that may include wonder, awe, and much sharing. It seems that the relationship of many noble environmentalists to the human body and to their own body borders on the abstract.

Consequently, if we carefully read the writings of many biologists and other scientists who strive to preserve biodiversity—among them Rachel Carson and Edward Wilson, whom we mentioned in previous chapters—we discover a disturbing fact. The statements and writings of these adherents of biodiversity may often convey the impression that their own body does not exist as a being that is living in the world and interacting with other beings with whom it shares that world. Very little of the wonder and awe that a person can bodily experience when watching a butterfly emerge from its cocoon, or the wonder and bodily joy that overcomes a person when encountering the beauty of a field of spring flowers swaying gently in the breeze—little of such wonder, awe, and joy is mentioned in their writings.

Here a short digression reveals the existential poverty that emerges in a person's life when he or she ignores one's body. Wonder, Plato asserted, is the beginning of philosophy. The search for truths, for knowledge, and for wisdom, he pointed out, must begin with wonder. We all know that wonder can be breathtaking in its literal meaning—it may take away one's breath. Hence, wonder can bring a person to stop his or her everyday busy-ness and comprehend one's own perceptions and field of perception that seem to be well known. Such comprehension may lead to thinking and to knowledge. But, an additional point should not be overlooked. When, say, encountering a beautiful field of flowers, my joy and wonder engulf my entire being. Thus, one major source of wonder and thinking is bodily involvement.

Furthermore, in the concluding pages of *Critique of Practical Reason*, Immanuel Kant expressed awe "at the starry heavens above me and the moral law within me."[2] Notice that Kant's awe is at what may seem to be everyday perceptions. We can assume that Kant's awe when encountering these simple perceptions—any person can see the starry heavens every unclouded night—contributed to his thinking.

There are yet additional elements of wisdom that a person who wonders may encounter and absorb. As Plato shows in the dialogue *Theaetetus*, wonder leads beyond understanding to comprehension of one's own limited knowledge and to the appreciation of the mystery of knowledge and truth. Such an appreciation of mystery, Socrates rightly intimated in the dialogue, adds depth and wisdom to a person's way of life. Many of these mysteries are bodily encountered by every person—for Kant two such mysteries were the starry heavens and the moral law within him. We would add that among such everyday mysteries you must include the mystery of learning and of love; and also the mystery of a butterfly emerging from a cocoon. Today we hold that wonder and awe, in this Platonic and Kantian sense, should be the beginning of a renewed bodily involvement with beings of nature.

Unfortunately, the awe and wonder that Plato and Kant described are rarely expressed by many of the scientists who are concerned with saving the environment. Much the same is true of the writings of many of those noble laypeople who struggle daily for the preservation of the countless species of nature that are endangered by the deeds of indifferent political regimes, by the human greed of the princes of corporate capitalism, and by the destructive actions that stem from their greed. We repeat: Many of these strugglers for preserving the species of nature do not express wonder or awe; many of them seem to relate to their own body as a passive anonymous being. You rarely find in their writings hints of the wonder and awe which bodily involvement with the beings of nature may bring forth.

An example of such non-bodily writing, in which one finds little wonder and awe, is the truth-clarifying and informative journal *Seedling*, which is published by the international non-governmental organization Grain—Genetic Resources Action International.[3] The dedicated writers in *Seedling*, and many writers in similar informative journals that reject the ruining of the environment, assume a supposedly objective perspective on what is happening in the world and especially in their writing about the destruction of natural species. One characteristic of such a so-called objective approach is the writers' attempt always to consider the topic that they are discussing from an overall objective perspective. However, by their reliance on this non-bodily perspective and overall view, these scientists and strugglers for the environment often detach themselves from their own personal bodily involvement in the world. Hence, they see no need to acknowledge or discuss their primal bodily relation to the other beings in the world, including those beings that they are endeavoring to preserve. The result is that their writings are intellectually informative, and reveal significant truths that the mainstream media conceals—but they frequently lack an existential dimension. Often, they also lack wonder, awe, and joy at the possibility of sharing this world with natural species.

Some of the sad outcomes of the fact that many environmentalists ignore the human body, including their own body, have already been indicated in previous chapters. The ignoring of the human body and its manners of perceiving and interacting in the world adds to the basic mistakes of these environmentalists that we already indicated—among them their reliance on the subject-object duality. These mistakes frequently lead many of these well-known environmentalists to embrace and promote an unexcusable philosophical superficiality. As indicated in previous chapters, we have sadly discovered that many of the strugglers for biodiversity only skim over the environmental problems that perturb them, without addressing the philosophical and existential roots of these problems.

Relying on the basic tenets of phenomenology, Merleau-Ponty states that human consciousness is anchored in and expressed by our bodily existence. As such, consciousness is expressed and involved in our daily bodily perceptions, responses, reactions, and engagements. Put differently, consciousness cannot be divorced from bodily engagements; consciousness is part and parcel of our everyday bodily involvement in the world. Furthermore, bodily engagements and bodily involvement are central to our perception of beings in the world, to the fields of perception that we establish and perceive, to the relations that we establish with the beings found in these fields of perception, and to our manner of existing in the world.

For instance, Merleau-Ponty argues that our sense and understanding of space is a result of our bodily engagement and bodily involvement. Our perception of space is primordially linked to our ability to move in the world during the engagements and involvements that we undertake. Consider a short citation from *Phenomenology of Perception* in which Merleau-Ponty presents ideas that describe the relation of our body to space: "the experience of our own body teaches us to embed space in existence . . . our body is not primarily *in* space; it is of it."[4]

We hold, but shall not attempt to prove, that the understanding of space that is presented in this citation, and in Merleau-Ponty's writings, improves upon Kant's thinking on space. This improvement emerges, we suggest, because Merleau-Ponty begins with the human body, and bases our grasping of space on our bodily existence.

In order to elucidate Merleau-Ponty's statements, we first want to point out that they rely upon, and also go beyond, Martin Heidegger's insights and thinking. Merleau-Ponty mentions Heidegger's thinking quite often in *Phenomenology of Perception*. Hence, before turning to Merleau-Ponty's discussion of space, we shall briefly outline Heidegger's insights concerning the term "being-in."

Heidegger explains in *Being and Time* that this term, being-in, which is part of Being-in-the-world, must be viewed ontologically.[5] He points out that a person is not *in* the world in the same manner that the tea is *in* the kettle, or the suit hangs *in* the closet, or the book can be found *in* the shelves of the library. The word "in" that is used in these phrases indicates that the items mentioned—the tea, the suit, the book—can be found in a specific place or within a certain receptacle. The phrases also infer that both the item mentioned and the receptacle exist in what we term "objective space."

Heidegger argues that, unlike these items that exist spatially in the world, a person is *in* the world as an engagement—he or she is in the world as a being that is always involved with and engaged with other beings. The fact that a person is spatially found in a room, a building, or a city, is secondary to the major manner by which a person is *in* the world, that is, by engaging beings in the world and being involved with those beings. Heidegger adds that, if a person's engagement and involvement in the world is authentic, it can lead him or her to thinking about Being, and to assuming responsibility and care for the beings that he or she encounters in the world. In certain instances, a person's authentic engagement and involvement in the world can bring forth important truths from unconcealment.

In the best of Heideggerian worlds, the human entity, that is in the world as involved and engaged with other beings, can fulfill one of the greatest challenges a person can undertake. This great challenge, Heidegger states, requires that the human entity assume the role of a shepherd of Being. A person who lives as a shepherd of Being cares for the beings that he or she encounters, and attempts to bring forth truths from concealment and think about Being. From this lofty perspective, Heidegger concludes that every person who is *in* the world participates in establishing the world. Hence, if there were more shepherds of Being, the world would be much better.

We cannot elucidate all the environmental implications arising from Heidegger's worthy challenge to human beings. Yet, even without an elucidation, one truth is evident. If human beings strive to act as shepherds of Being, and not as exploiters of beings, there will be great benefits for other natural species and for other beings with whom we human beings share the world.

It is clear that in his discussion of perception and of our relationship to space, Merleau-Ponty accepts Heidegger's ideas that a person is *in* the world as a being who is engaged in the world and involved with other beings in the world. With this fact in mind, we can look again at the citation: "the experience of our own body teaches us to embed space in existence . . . our body is not primarily *in* space; it is of it." The citation shows that, in relation to space, Merleau-Ponty proceeds further along the path opened by Heidegger's thinking. Yet a word of caution is necessary. Merleau-Ponty's thinking does not proceed in the direction of the superficial pantheistic mysticism, attributed to his writings by thinkers such as David Abram. Nor do Merleau-Ponty's writings have any relevance for the weird field of "deep ecology," toward which Abram attempts to lead his readers.[6] We often wonder: Have Abram and other "deep ecologists" never heard of a philosopher called Friedrich Nietzsche? We ask because, if the "deep ecologists" would have read Nietzsche's ideas about the futility and stupidity of idealizing nature, and learned from him, they would have seen the inanity of presenting and romantically idealizing a so-called realm that they called deep ecology.

One reason for Merleau-Ponty's proceeding further than Heidegger is found in his emphasis on the body as a phenomenological source. This emphasis on learning from the human body, and especially as a body that moves in the world as a Being-in-the-world, teaches Merleau-Ponty that the human body is of space. To the best of our knowledge, Heidegger did not elucidate in any detail or depth a person's bodily relationship to movement and to space. Merleau-Ponty deals at length with movement, and with the ways our body moves in the world. He shows that bodily movement is central to perception and to the fact that space is embedded in a person's existence. Merleau-Ponty includes in this relation to space the grasping of depth. He claims to have found failings in the writings of George Berkeley, who discussed our perception of depth at length. We shall not elucidate Merleau-Ponty's argument and findings on this matter.

Yet, someone may ask: What exactly does Merleau-Ponty mean when he states that space is embedded in a person's bodily existence? What does he mean when he holds that our body is of space?

In the most simple terms, Merleau-Ponty means that my body is always and at every moment living as a spatial being and relating spatially to other beings. Every color and form that I see is in space, every word that I hear comes from somewhere in space, as does every odor that I smell. All movements, all interactions of my body require that I live as a spatial being. At this moment, I am living as a spatial being and relating spatially to the cup of tea that I am raising to my lips. After drinking a few gulps of tea and setting down the cup on the table, I will relate spatially to the keyboard of my personal computer when I press the keys so as to continue writing this book. Our body is of space because any human perception or action or movement in the world requires that the person's body relate spatially to other beings.

Consider the sad example of a person who is blind and must walk and find his or her way in the world with a stick. Merleau-Ponty describes in well-known details the blind person's use of the stick; but he views this finding of a way in the world from the perspective of the body's relation to space. In order to move, the blind person's body must act spatially in the world. The blind person's stick serves as an extension of his or her fingers and as an assistance to the task of walking and spatially comprehending the world. Put differently, together with the process of assisting the blind person find his or her way while walking, the stick helps the blind person relate spatially to the beings that he or she encounters, and to the world. Thus, through his or her own bodily movements and the gentle manipulating of the stick in space, the blind person relates spatially to other beings without seeing the other beings that exist in the world. We again perceive that the body—in this case the hand holding the stick and the blind person's feet that follow the stick—is essential to these movements and to establishing spatial relations in the world.

Someone may still ask: What are my bodily relations to space during moments of relaxation, say, when I rest quietly, without moving my body? For a novelist's answer to the question, we turn again to Marcel Proust's *Remembrance of Things Past*.[7] Proust's descriptions of relaxation and rest constitute an answer that fully accords with Merleau-Ponty's statements which articulately explain that every sensation is spatial.[8]

In the opening pages of *Remembrance of Things Past,* Proust describes in revealing detail the problems his narrator, Marcel, encounters when attempting to fall asleep; he also describes the bodily situation in which Marcel finds himself when awakening from a deep sleep. Marcel describes these problems and situations during different stages of his life, and also when he slept in different bedrooms, in different beds. Proust shows that, even in those moments when I lie relaxed in my bed, free of tasks and waiting to fall asleep, or after waking from a deep sleep, my body is spatially involved in the world and relates spatially to other beings. Such occurs because every sensation is spatial. Thus, I may lie in bed and feel the softness of my pillow on my cheek, or gaze at the sliver of light that can be discerned under the door, or hear the slow toll of the village clock announcing the hour. Frequently, Proust suggests, these sensations are tinged with bodily remembrance of things past; at times the spatial relations that I perceive seem to be born of sensation-inspired memories. Yet, what unites all these situations is that Marcel relates bodily to his immediate environment. Proust's novel clearly shows that, as Merleau-Ponty stated, every sensation is spatial. Thus, even when resting, space is embedded in my existence and my body is primarily of space. In all moments of my life, my body exists as a spatial being which consistently relates spatially to other beings.

Three immediate conclusions emerge from Merleau-Ponty's presentation of the spatial existence of the human body. First, the existing human body that perceives, moves, and acts in the world is the primal and original source of every person's living in space, perception of space, and conception of space. Second, our understanding of objective space emerges on the basis of our perceptions, our bodily movements, and our bodily engagements with the beings that we encounter. Furthermore, since a person's body is of space, living in space precedes any abstract conception of space. And finally, as a corollary, all human perception and interaction in the world is a bodily and spatial involvement. These conclusions indicate that in relation to space, as in relation to many other matters, we should relate to the human body as a phenomenological source of knowledge and understanding.

Merleau-Ponty's poignant and enlightening formulations—that space is embedded in our bodily existence and that our body is of space—hint at the breadth and depth of his study of the human body and its relationship to the world. But like many an original thinker, in presenting his findings, he is frequently confined by the terms of accepted language and by the rules and concepts prominent in scientific discourse. In coping with this difficulty, Merleau-Ponty does not act as did Martin Heidegger and Martin Buber, who coined new terms such as "Dasein" and the "I-Thou" in order to present novel ideas that are central to their ontology. In imparting his thinking, Merleau-Ponty traverses a different path. At times, when he encounters the difficulties of expressing his thoughts within the confines of language, he presents vivid poetic analogies. The role of these poetic analogies is to clarify. Evidently, Merleau-Ponty believes that the poetic analogies explain his ideas and illuminate significant details of the ontology of perception that he presents.

Here is a short example of such poetic analogy, presented earlier and discussed briefly in Chapter 2. We return to it frequently in this book because of its centrality to Merleau-Ponty's discussion of the human body and its relation to the world.

Our own body is in the world as the heart is in the organism: it keeps the visible spectacle constantly alive, it breathes life into it and sustains it inwardly, and with it forms a system.[9]

Even before partially presenting Merleau-Ponty's descriptions and arguments that support this citation, consider his major point. We must relate to the human body as existing in the world in the same way that the heart exists in the organism. Now, if Merleau-Ponty's poetic analogy unconceals an important truth, some conclusions are forthcoming. One immediate conclusion arises from the fact that we briefly mentioned in Chapter 2, but is worth repeating: The heart of an organism will never deliberately act in order to ruthlessly destroy the organism in which it resides. But, such does occur with many human beings who have bodies and who currently exist in the world.

Yes, we are repeatedly faced with the fact that many human beings are ruthlessly destroying the natural environment and the world that they share with other beings. Yet, according to Merleau-Ponty, each of the bodies of these destroyers of the world, bodily resides in the world as a heart. Consequently, each person who initiates rampant destruction of the natural world—and today there are many many thousands of such people, especially among those employed by certain ruthless capitalist corporations—resembles a heart that betrays and destroys its own body.

Merleau-Ponty's poetic analogy concerning the human body unconceals additional significant truths which have relevance to our relationship to nature. It discloses an important truth that accords with the tenet that human consciousness is anchored in and expressed by bodily existence. To see this truth, consider a simple example that he gives of my body being in the world as the heart is in the organism.

Merleau-Ponty points to the fact that I am bodily acquainted with the apartment in which I live. Moreover, this acquaintance is never abstract. It is not a result of my perceiving the apartment from different perspectives and creating a mental map of it. Nor is my acquaintance with my apartment a result of my calling up a mental bird's-eye view of it. My body lives

The Human Body as a Phenomenological Source

in this space and has what may be termed "the feel of the apartment"—without my needing to abstract, analyze, reckon, or calculate anything that has to do with my involvement in the apartment. For my body, my apartment is not primarily a walled-in portion of space that has been set off from the other apartments in the building, and in which can be found furniture, utensils, music disks, my personal belongings, my desk, and my books. My body participates in converting this space, this furniture, these personal belongings into a space that it recognizes as my home. Hence, my body is a major phenomenological source of my being at home in this apartment. This statement can be justified by a simple test: If I feel bodily alienated in the apartment and cannot find my way around it, the apartment is no longer a home. Before going further, we can ask a rhetorical question: How many people today feel at home in a section of the natural world?

In order to further elucidate the manner by which I feel bodily at home in a certain space, we borrow two of Merleau-Ponty's vivid insights. The first insight concerns a woman with a tall feather in her hat. We hold that in my acquaintance with, living, and feeling at home in my apartment, I resemble the woman with a tall feather in her hat that he describes. Consider the woman. Without calculating or reckoning, Merleau-Ponty notes, the woman usually knows exactly how much she must bend her head in order to pass through any door that she encounters—so that the feather will remain intact and unharmed. Seemingly, the feather is an extension of her head.

In another example, Merleau-Ponty points out that the experienced driver bodily senses if there is enough room for the car he or she is driving to pass through a partially open gate without scratching its side. Neither the woman nor the driver measures distances; they react bodily to the problem that confronts them. Yes, like the feather in the woman's hat, the car is seemingly an extension of the driver's bodily involvement in the world. Put differently, by extending its involvement in the world to the material objects with which it is engaged, such as the hat with the tall feather or the car, my body seemingly converts these material objects into extensions of its own being in the world.

Similarly, I feel at home in my apartment through hundreds of details of everyday life in which I am bodily involved in the apartment, and in which my body is extended beyond its borders through this involvement. At times, these details can be minute or seemingly insignificant—for instance, I usually know exactly how much to turn my head or body so as to see what time it is on the kitchen clock. These bodily involvements breathe my personal life into this specific walled-in portion of the world. And this continual bodily engagement in this small walled-in area in which my belongings are found, and which I endow with significance for my existence, makes this small area my apartment. Thus, my bodily involvement in my apartment, that again and again occurs during my daily life in the apartment, "keeps the visible spectacle" of my apartment "constantly alive."

Again we can note that few people today feel at home in specific areas of nature in the manner that I may feel at home in my apartment. Such need not be. As Mark Twain showed in *Adventures of Huckleberry Finn*, Huck Finn and Tom Sawyer grew up to feel at home in nature—much as many children today are at home in a city apartment. We believe that those children who feel at home in nature—and we acknowledge that there are not many—will not view the beings of nature that they encounter merely as resources that can be exploited for their personal benefit.

Merleau-Ponty's cited poetic analogy is instructive for our relationship to the environment not only in the manner that we previously articulated—as a condemnation of those greedy people and corporations who destroy the face of the earth and eradicate many species of living organisms. Since the poetic analogy is based on an ontology of sharing the world, it can also suggest a much better way of living in the world. Such a way of life would reject the manners of existing in the world proffered by the promoters of corporate capitalism or the idealogues of other materialistic approaches which prevailed in the twentieth century.

We have repeatedly suggested that the search for a way of life of sharing is most important, given the well-known widespread devastation wreaked upon the face of the earth by the supporters and upholders of the dominant regimes in the twentieth century. As we have indicated, these regimes related to people and to other living beings, at best, as mere resources or commodities. Often in these regimes, people and other beings that are found in the world were considered to be superfluous. Research in the social sciences and original works of art reveal that many people who reside in the shantytowns of South America or in the black ghettos in American cities grasp that they are considered to be superfluous. Hence, it is not surprising that capitalism, communism, and all of the other materialistic approaches of the past and present resulted in killing many human beings.

Furthermore, if millions of human beings are considered to be superfluous, or, at best, to be a resource or a commodity, the attitude toward natural beings can hardly be better. The result of indifference toward fellow human beings and toward natural species in the past half century is well documented. We have witnessed rampant destruction of large areas of forests, wetlands, prairie, rivers, the oceans, and many other parts of our natural environment. Learning from this short sad history, we hold that indifference and destructive attitudes toward human beings and toward beings of nature may lead to a catastrophe. Indeed, the situation today is very discouraging. Those regimes who consider human beings to be mere resources continue to thrive and to make human beings superfluous, to convert large portions of the earth into a wasteland, and to kill off many living species. Need we add that such destructive approaches are spiritually barren?

As against these spiritually barren approaches, a reviving fresh breeze of sharing the world with other species emerges from Merleau-Ponty's poetic analogy. Look again at the locution: "our body is in the world as the heart is in the organism." Perhaps we should stretch Merleau-Ponty's poetic analogy a bit and state that, when a person's body acts like a good heart, it can bring about a better world. Indeed, it is evident that a person who bodily respects and cares for other beings may bring some good to that portion of the world in which he or she resides and acts. Using Merleau-Ponty's language we can add that, at times, such a person may breathe life into the beings of the world that he or she encounters. Need we add that such a breathing of life into the beings that a person encounters in the world is the opposite of wrecking destruction upon those beings?

Furthermore, an organism's heart should live in harmony with the organism. We recognize that living in harmony in our world, in which corporate capitalism reigns and in which it daily extols greed, is no simple task. Still, we believe that concerned and sensitive human beings can attempt to reject these ruinous capitalist approaches and endeavor to live in some sort of bodily harmony with the beings that they daily encounter in the world. At least, a person can endeavor to relate with respect to other natural beings that he or she encounters in the world. Santiago in the citation from Hemingway's *The Old Man and the Sea* in Chapter 1 expresses such respect for the large fish whom he struggles to capture and kill.

Whether they know it or not, persons who endeavor to live in harmony with beings of nature are relating to their own body as a phenomenological source. Look again at Santiago in *The Old Man and the Sea* and this truth will emerge. He speaks to the fish and admires it. Santiago even speaks to his hands. This truth also emerges in Hemingway's autobiographical essay *Green Hills of Africa*.[10] These books by Hemingway, and the writings of other great authors such as Tolstoy, reveal that the possibility of a person endeavoring to live in bodily harmony with the beings of nature is not far fetched.

Perhaps it is wiser to suggest that the idea of living in bodily harmony with the beings of nature is often considered to be far fetched by many—if not most—of the city-dwellers who inhabit the world. We can include those affluent families who populate the suburbs of the cities among the city-dwellers. The daily life of these city-dwellers, among whom you may find many members of the intellectual elite, is frequently alienated from the remarkable and subtle manifestations of nature. Such people rarely experience the engaging encounters that a person may have with the beings of nature. Thus, city-dwellers may reside, as some who have homes in the suburbs do, in a house surrounded by a flower garden, a few trees, and a manicured lawn. But many—if not most—of the residents of such homes have few unplanned bodily encounters with beings of nature. Nor will most of these city-dwellers relate with Platonic wonder or with Kantian awe when encountering some of the simple manifestations of life in nature.

What is more, many of these city-dwellers are alienated from what is occurring in farms and in the realm of farming all over the world. We suspect that many of these city-dwellers do not imagine or comprehend that farmers can strive to live in bodily harmony with natural beings that exist in the world, and especially in the area that they farm. But the truth is that today you can still find many millions of farmers who attempt to live in bodily harmony with the beings existing in the world, and among these beings are domesticated animals. These attempts to live in harmony are prominent among the many millions of what may be called "traditional farmers"—those farmers who daily work their land in traditional ways.

By "traditional farmers" we primarily mean those millions of farmers in Third-World countries in Africa, Asia, and South America, who, as yet, have not been forced to accept the corporate capitalist industrialization of agriculture. Nor have these farmers, as of yet, been forced to accept the cult and methods of intensive farming that accompany the capitalist industrialization of farming. For anyone who has seen an industrialized chicken egg-laying farm, in which the caged chickens can hardly move about and never see the light of day, it is not difficult to understand what is meant by "the industrialization of farming." Yet, we want to cite an enlightening remark about industrialization of agriculture by Jose Bove, a French farmer who has struggled for many years against intensive farming and its destructive results. We should perhaps add that Bove has been jailed a few times for his struggle against the ruining of the natural world by agricultural corporations and other companies that promote intensive farming and industrial agriculture. Bove has also struggled against those companies who wish to spread genetically modified plants that may ruin the environment.

Bove's statement discloses one of the major dangers to our relationship to the soil and to the land—a danger that has been brought about by industrial farming. Bove states: "In intensive farming the object is to adopt the soil to the crop, never the other way round. This is a fundamental change in agriculture."[11] In short, in intensive farming, the soil that is farmed is a resource that is to be manipulated—the soil is not lived with, not related to, not respected as the source from which living beings draw sustenance.

Today, the non-intensive environmentally sensitive farming methods of the traditional farmers in the Third World are endangered. Consider the Green Revolution, that was lauded and applauded by all corporate capitalist leaders, their sycophants, and many scientists in the 1960s and 1970s. With the scant wisdom that comes with hindsight, we now know that the Green Revolution brought few, if any, long-lasting benefits. Over the years, and with the many failures that it brought, we have learned to see the Green Revolution as an attack on traditional farming and as a source of profits for a small group of corporate capitalists.

Yes, we now know that the Green Revolution was a major attempt by capitalist corporations, working together with government bureaucrats who had little understanding of the environment, to change traditional farming approaches in the Third World in favor of intensive farming. That goal was partially achieved, and the corporations involved in the revolution reaped great profits. The indigenous farmers, say, in India or Mexico, reaped much less. Frequently, many of the farmers lost a source of livelihood and a way of communal life. As Vandana Shiva has shown in enlightening detail, the Green Revolution brought about the widespread ruining of agricultural land, the polluting of water supplies, the destruction of humanly supportive social structures, and the devastation of the natural environment.[12] Together with many others, Shiva is struggling to help some of the farmers who suffered from the ruin spread by the Green Revolution return to their traditional methods of farming.

We should also mention that there exists a large new group of "traditional farmers." They are the many thousands of farmers of Europe and North America who have rebelled against the methods of intensive farming and its destructive results for human beings and for the environment. (Jose Bove is a member of this group of farmers.) These farmers reject the trend central to intensive farming that relegates both human beings and the beings of nature to the status of mere resources and commodities. These new traditional farmers also reject any farming whose primary goal is to greedily exploit the land or any other living species. Many of these First-World farmers have returned to traditional methods of farming, such as crop rotation and sending the cows out to graze in the pasture. Some of them have also promoted what is called organic farming. Evidently, in each farmer's decision to use traditional methods in farming, his or her body is a phenomenological source for establishing a relationship of respect toward the land that one cultivates and toward other natural beings.

Both groups of traditional farmers, we believe, tend to relate to the land that they farm as a partner in one's worldly endeavors. They view the land that they cultivate, and the beings which reside upon the land, as worthy of respect. Hence they wish to live in some sort of bodily harmony with these beings with whom they share the world.

Let us repeat this point. Land and soil, for these farmers, is not merely a resource that exists in order to be exploited by human beings or by large conglomerates and multinational companies. Land and the soil are a source of all existence, and a partner to all worldly existence, including human bodily existence. Hence, many of these farmers strive to be in some sort of bodily harmony with the land which they cultivate and the beings that they encounter while farming.

In summary, we should add another major point arising from Merleau-Ponty's discussion of the body. He indicates that living one's body as a phenomenological source is very much an expression of freedom. We believe this view to be true and most significant for human existence. We already mentioned love, learning, wonder, awe, and joy, which are unique expressions of human freedom.

We can add that many farmers who wish to live in harmony with the beings of nature, and who engage in methods of traditional farming, such as Jose Bove and his comrades, would probably agree that living fully with one's body is crucial for human freedom. With this insight in mind, we turn to Part Two, in which we discuss the link between Merleau-Ponty's phenomenology of perception, freedom, and the significance of fields of perception for sharing the world with other beings.

NOTES

1. See Maurice Merleau-Ponty, *Phenomenology of Perception,* trans. Colin Smith (London: Routledge & Kegan Paul, 1962), 164.

2. Immanuel Kant, *Critique of Practical Reason,* trans. Lewis White Beck (Indianapolis, IN: Bobbs-Merrill Co., 1956), 166.

3. *Seedling* is published by Genetic Resources Action International whose offices are in Barcelona, Spain. The journal describes the environmental destruction brought by genetically modified crops. For instance, its January 2003 issue includes enlightening essays on the ruining effect of the genetically modified Bt cotton and on the genetic pollution of maize.

4. See Merleau-Ponty, *Phenomenology of Perception,* 148.

5. Martin Heidegger, *Being and Time,* trans. John Macquarrie & Edward Robinson (Oxford: Basil Blackwell, 1962).

6. David Abram, "Merleau-Ponty and the Voice of the Earth," in David Macauley (ed), *Minding Nature: The Philosophers of Ecology* (New York: The Guilford Press, 1996), 82–101.

7. Marcel Proust, *Remembrance of Things Past,* trans. C. K. Scott Moncrieff (New York: Random House, 1934).

8. See, for instance, the discussion of sleep in *Phenomenology of Perception,* 164. See also 221.

9. Ibid., 203.

10. Ernest Hemingway, *Green Hills of Africa* (London: Arrow Books, 1994).

11. Jose Bove and Francois Dufour, interviewed by Gilles Luneau, *The World Is not for Sale: Farmers Against Junk Food,* trans. Anna de Casparis (London: Verso, 2001), 67.

12. For more on this topic, see Vandana Shiva, *The Violence of the Green Revolution* (London: Zed Books, 1991); Vandana Shiva, *Stolen Harvest: The Hijacking of the Global Food Supply* (London: Zed Books, 2000).

Part Two

Freedom and the Field of Perception

5

The Human Body and the Field of Perception

In *Phenomenology of Perception,* "the field," that is, the field of perception, is a concept that is prominent in Merleau-Ponty's discussion. He explains that "any sensation belongs to a certain *field.*"[1] Sensations (or sense-data, as philosophical analysts call them) are not out there, waiting for us to perceive them. Rather, Merleau-Ponty stresses, sensations address us as part of a field of perception to which we contribute our being. Consequently, when we attempt to elucidate the phenomenology of perception, the field of perception is a key concept.

Merleau-Ponty does not give a formal definition of the field of perception. Yet, throughout *Phenomenology of Perception,* he does describe the manifestations of the field of perception that human beings establish. He also points out the centrality of the field for perception, and the implications arising from the fact that human beings have fields of perception. These fields of perception are personal and are linked to each person's bodily existence. They are partially an outcome of the intentions of the person who perceives and who establishes the field of perception. Specific fields of perception will frequently be a result of chance or of educational and cultural influences. Consider, for instance, a daughter of a composer and conductor of classical music. There is a much greater chance that she—and not the son of a poor tinker—will establish a field of auditory perception in which the music of Bach is prominent.

In what follows, we shall not describe all of Merleau-Ponty's descriptions and discussions of the field of perception. Our undertaking is modest.

With the help of some of Merleau-Ponty's insights, we shall endeavor to partially clarify the concept of the field of perception, and its links to the human body. We shall also describe the fact that the field of perception is not distinct from the body of the perceiver. We shall point out that this fact has significant philosophical implications. One implication, we believe, is a firm rejection of individualism as a creed, and the promoting of a way of life in which persons share the world with other beings.

Throughout the *Phenomenology of Perception,* Merleau-Ponty uses the concept of the field. Yet, only on page 328 does he present a statement that can serve as a partial definition: "The field is a setting that I possess for a certain type of experiences, and which, once established, cannot be nullified."[2] The first part of this definition seems evident. We know that our visual perceptions occur within a visual setting, in which light is most significant. We also know that hearing occurs within an auditory setting, say, in which there is a medium that conveys the sound waves. If the auditory setting is beset by stark interferences, hearing becomes problematic. We cannot hear a person's voice when it is drowned out by thunder.

Yet, what does Merleau-Ponty mean when he states that the field of perception cannot be nullified? We shall answer with the help of two well-known historical examples in which a person's field was not nullified even though the sense which had helped to establish the perceptual field was damaged beyond repair. Each example helps to clarify what it means to possess a setting for a "certain type of experiences." It also shows the profound significance of a field of perception for human interaction, creativity, and knowledge.

The historical examples are Ludwig van Beethoven's auditory musical field and John Milton's visual field. We agree that their fields were exceptional, they are the fields of geniuses. Yet, for the theme of this book, Beethoven's musical genius and Milton's poetic genius is secondary to the ontological descriptions that Merleau-Ponty presents. We shall also suggest that Merleau-Ponty's description of the field of perception accords with a well-known biblical story.

By relying on his auditory musical field, Beethoven created original musical works of lasting beauty. True to Merleau-Ponty's statement, this auditory field was not nullified when Beethoven became totally deaf at the age of 49, in 1819. After his affliction, communication with him was possible only in writing. It is a historical fact that Beethoven's being deaf did not halt his composing of beautiful music. In the language of Merleau-Ponty, thanks to the existence of his musical field—this auditory setting which Beethoven possessed and which was not nullified by his deafness—Beethoven continued to compose magnificent musical works of art. Thus, for at least seven years, until his death in 1827, Beethoven wrote beautiful musical works—even though he could no longer hear the musical notes that he wrote.

Look, for instance, at the history of Beethoven composing his ninth symphony, which is considered to be one of his magnificent musical achievements.

Beethoven commenced writing the symphony in 1817; he completed it in 1824. Consequently, Beethoven composed large parts of his ninth symphony during the years that he was totally deaf.[3] Learning from Merleau-Ponty, it is correct to state that the deaf Beethoven finished composing his ninth symphony on the basis of his not-nullified field of auditory perception.

Someone may ask: What is the relationship between a specific sense and its field of perception? Merleau-Ponty points out that a dynamical relationship exists between a certain sense, say, seeing or hearing or smelling, and the field or fields in which that sense is engaged. The sense is necessary for establishing the perceptual field; but once the perceptual field is established, the sense is subordinated to it.

If Beethoven had been deaf from birth, he could not have possessed an auditory field, which included musical notes. He could not have established and possessed a setting for his broad and profound auditory experiences. His musical genius would never have come into being, and he never would have composed music. Beethoven's sense of hearing was necessary for the establishment of his auditory field, and for the development of his musical creativity. But once his auditory musical field was established, Beethoven's sense of hearing was "subordinated to a certain field."[4] One such field was the specific musical field that he had learned to hear and later possessed.

We have already stated that much of what is learned from the example of Beethoven and his auditory field is true of other sensual fields of perception. Thus, a person who is born blind has no access to a visual field. He or she will pass through life without such a field. In contrast, quite frequently persons whose vision has become impaired, or who have become blind during their life, still have access to their visual field. The well-known biblical story of the blind Isaac is a case in point.

Isaac wanted to bless his first-born twin son, Esau. Because of his blindness, however, Isaac blessed his son, Jacob, whom he believed to be Esau. Isaac's blessing of Jacob occurred because Isaac's wife, Rebekah, and Jacob took advantage of Isaac's blindness and deceived him (Genesis, 27). Rebekah sent Jacob, dressed up as Esau, to receive the fatherly blessing. In the biblical story, Isaac was unable to see in order to determine which of his twin sons he was blessing. But, and here we note the linkage to Merleau-Ponty's statement, the wording of Isaac's blessing to Jacob required that he have access to a visual field. In his blessing, Isaac mentioned bounty that is seen. He promised Jacob that he would enjoy the dew of the skies and the fats of the earth, with much corn and juice of the grape. Isaac's blessing is definitely that of a person who has access to a field of vision. No person who is blind from birth can speak intelligently, on the basis of experience, about the sky.

We grant that the biblical story of Isaac has no historical evidence to support it. But much the same conclusions concerning a visual field of perception can be inferred from the life of John Milton (1608–74). Beginning in 1656, when he was totally blind, Milton wrote *Paradise Lost*, which many scholars consider to be his major contribution to English literature.

Milton dictated verses of this book-length poem daily to a secretary and completed the volume-poem in 1663. His visual field of perception remained vibrant, despite his blindness; hence he could dictate sentences that are based on a visual perceptual field, such as the following:

> I thither went
> With unexperienced thought, and laid me down
> On the green bank, to look into the clear
> smooth lake, that to me seemed another sky.[5]

We soon turn to Merleau-Ponty's assertion that human freedom is firmly linked to our fields of perception. Such is already evident in the three examples. It was within his musical auditory field, within the specific musical setting which he possessed, that Beethoven, at least partially, lived his freedom as a composer. This freedom was necessary for Beethoven's musical genius to find a way to reach expression. The results are well known. By living his freedom creatively within his musical field, Beethoven brought into being musical compositions whose magnificence endures, and whose beauty may enhance the being of their listeners. Much the same can be said of John Milton and the creation of *Paradise Lost* and of other poems that he wrote after becoming totally blind. Milton lived his freedom within his visual field of perception which remained vivid, despite his blindness.

Similarly, Isaac lived his freedom in deciding to bless Jacob, whom he believed to be Esau. The Bible describes Isaac's hesitation, which means that he was free to not bless Jacob.

As mentioned, a major point that emerges in *Phenomenology of Perception* is that a person bodily lives his or her freedom within a field of perception. The human body is a participant in the field of perception, which a person perceives and in which the person is engaged. Put differently, a person's body is always involved in, and constitutes a major component of, each of his or her fields of perception. Thus, there is not a field of perception out there, and a person who contemplates the field over here, detached from the field. Rather the person's body is intimately and firmly involved and engaged in the field of perception which the person perceives. A person's body shares its being with every one of his or her fields of perception.

We now can see a new depth in Merleau-Ponty's poetic statement that we quoted in preceding chapters:

Our body is in the world as the heart is in the organism: it keeps the visible spectacle constantly alive, it breathes life into it and sustains it inwardly, and with it forms a system.[6]

The statement teaches us that each field of perception in which the human body is involved is part of the system that the body forms.

A person's body breathes life into the system and sustains the fields of perception in which that person is involved.

One way that Merleau-Ponty enlarges upon and elucidates the body's involvement in the field of perception is by relying on experiments published by certain psychologists of perception. These psychologists discovered, Merleau-Ponty shows, that the field of perception can seemingly correct itself when faced with grating or disturbing perceptual problems. Merleau-Ponty discusses two such experiments: when a person is put into a room in which mirrors distort spatial relations, and when a person walks around wearing eyeglasses that invert his or her visual field. The experiments revealed that, at first, the person undergoing the experiment suffered from disorientation in his everyday movements and actions. After a period of time, however, the person's field of perception corrected itself and he or she could move about freely in the room, pick up objects, and use them intelligently. Such occurs, Merleau-Ponty explains, because the person's body lives in the room as it is distortedly reflected by the mirrors or by the eyeglasses. The engagement of living in the room requires correcting the field of perception, a task that the body undertakes. For this correction of the visual field to occur, reflection is not needed. Put succinctly, in these experiments, the person's body is engaged in perceiving the objects in the room and the room itself, and is involved in correcting the distorted visual field.

Merleau-Ponty's concepts, with which he describes a person's involvement with fields of perception, provide additional insights as to the relationship of the body to the field of perception. Thus, the establishing a field of perception, and in many instances the ability to correct that field, Merleau-Ponty explains, is partially a result of the human body being an "expressive unity."[7] As an expressive unity, the human body is unique, and this uniqueness brings forth the field of perception and, as Merleau-Ponty notes, breathes life into it. Put succinctly, the "expressive" bodily existence that is unique to human beings is an ontological source of the establishment of fields of perception. Our continual bodily involvement in these fields of perception is also based, of course, on the senses, but also on the human body being an expressive unity. Here is one additional example.

We wish to consider what happens, at times, when some people listen to a superb concert, say, when a great pianist accompanied by an accomplished orchestra perform Rachmaninov's Third Concerto for Piano and Orchestra. Suddenly, in unique moments during the concert, when the pianist reaches heights of excellence and expression, these people will feel their entire body responding, their skin may bristle, or their breathing may be held or become irregular. What has happened in these unique instances is that each listener's body is totally involved in the auditory field of perception, and it is responding to the concert as an expressive unity.

The situation becomes all the more complex when we add to these fields of perception an additional crucial area in which our body is a unique expressive unity: the hearing and articulation of language. We shall not expand on this theme, but an anecdote may be of help in hinting at the complexity of our bodily living with language.

It is well known that many Italians "speak with their hands." The story is told of a key Italian spymaster who was captured by the enemy. For weeks they severely tortured the spymaster, demanding that he tell all that he knew. Despite his great suffering, the spymaster never disclosed a word. When he finally was liberated by Italian forces, friends asked him: "How could you stand up to all those long hours of torture and not speak a word?" "It was not that difficult," the spymaster answered. "Whenever the enemy interrogated and tortured me they always tied my hands. With my hands tied, I could not speak, even if I wanted to!"

In describing the body as an expressive unity, Merleau-Ponty is going beyond many of the philosophical approaches that prevailed during long historical periods. An exception to this trend was Friedrich Nietzsche, who also grasped the body as an expressive unity, and urged thinkers to be attuned to their bodies. Yet, to the best of our knowledge, Nietzsche did not use the term "expressive unity." As indicated in previous chapters, many past and some contemporary philosophers relate to the body abstractly, as a passive and anonymous being. Unlike Nietzsche, and unlike many existentialist philosophers who do discuss our bodily involvement, these thinkers consider the human body to be not much of a consequential being. They concede that the body is somehow attached to the mind or the soul—which is the source of thinking and of human dignity and integrity. Many of these thinkers never imagined the possibility of relating to the human body as an expressive unity and as central to a life of freedom.

Together with Merleau-Ponty and Nietzsche, we hold the accepted philosophical approaches which ignore the human body to be ontologically flawed. Those thinkers who blatantly disregard the human body as an expressive unity that helps to establish fields of perception, and is involved in these fields, purposely overlook and suppress an entire dimension of human existence. Furthermore, they do not comprehend how the human body, as an expressive unity, is involved and engaged with other beings in the world. They do not perceive that the human body as an expressive unity can establish and possess settings in which certain types of experiences are prominent. They do not grasp that these settings of types of experiences, these fields of perception, will often greatly enhance human existence.

Merleau-Ponty's ontological assertion that any field of perception includes my body, which "breathes life" into that field, has many enlightening implications. We shall discuss only a few implications that are pertinent to the theme of this book. One of the most significant implications for our study is that, on the primal level of perception, every person shares his or her body and being with other beings that exist in the world. Without such sharing of one's being with other beings, a person could not establish and perceive a field of perception; nor could a person relate to the specific beings that exist in that field.

Hold it! someone may say. What happens at birth? It is unlikely that at birth babies are thrown into a field of perception and that they bodily share this field with other beings.

We do not know the answer to this question. To the best of our knowledge, this question was not addressed by Merleau-Ponty. Nor have we found answers in the writings of scholars who studied Merleau-Ponty's writings. Even those scholars who mentioned infancy do not address this question.[8] We would tend to believe that in the period before the newborn baby shares its being with other beings and establishes its fields of perception, he or she lives in what William James called "one big blooming buzzing Confusion."[9]

Let us be succinct. In *Phenomenology of Perception,* Merleau-Ponty does not attempt to explain in what manner the newborn baby begins to share its being and establish fields of perception. He does not relate to the "big blooming buzzing Confusion" that James assumed is the baby's fate in its early life. Nor, to the best of our knowledge, does Merleau-Ponty answer the question in other writings.

Yet, the fact that we cannot explain the manner by which perceptual fields come into being during early childhood does not diminish the centrality of the field of perception in human existence. We assume that certain fields of perception slowly replace the baby's "big blooming buzzing Confusion" that James presumed. We also know that, as the child grows and develops toward adulthood, it is incorporated into a world that it shares with other persons; in the early months of the baby's life, it shares the world primarily with its mother. That shared world, for each person, is constituted by fields of perception.

Some of the political, environmental, and economic implications of our bodily sharing the field of perception in which we exist have been mentioned in previous chapters. We believe that one of the major implications of our bodily sharing the field of perception counters a creed that daily invades our lives, and frequently governs our being. This creed has a rich history. Today the creed dominates much human existence, probably because it supports those people who currently wield power in large sections of the world. It is a creed that justifies widespread poverty, injustice, and much environmental destruction. It is a creed that fits contemporary capitalism perfectly. We are thinking of the individualistic creed.

Even in a sketchy overview, you will soon discover that the history of individualism is not necessarily a pretty history, especially in its recent manifestations. Yet, over the past centuries some of the finest minds have promoted and supported individualism. Consider some of the philosophical and scientific highlights in the history of thinking that supported the individualistic creed. These writings and scientific findings extend from Hobbes's *Leviathan* in the seventeenth century, through Benthamite utilitarianism, and up to individualism's nineteenth-century flirt with liberalism—in the Victorian utilitarianism of John Stewart Mill and others. For instance, John Stewart Mill, who is still considered an icon of personal liberty, supported the so-called principle of free trade, and with it, the Opium Wars against China. Included in this history of individualism are Darwin's scientific writings concerning the origin of the species and the survival of the fittest. Finally, in the late twentieth century and early twenty-first century, you find the neo-Darwinist, who finds genetic facts to support a radical individualism; in close neighborhood to the neo-Darwinist reside the economic neo-liberalists.

What are the messages of these contemporary individualists? Adherents of the first group, the neo-Darwinists, blatantly suggest that individualism is genetically innate and, hence, our fate. Each gene is a sort of individual that strives for its success in future generations. Members of the neo-liberalist group announce that, for society, the spread of individualism is an economic blessing.

We should add that the neo-Darwinism that currently prevails abounds with scientific flaws. There is no proof at all for its great belief in individualistic genes competing for the survival of the fittest.[10] Also, we have already indicated that neo-liberalism, in addition to many other evident evils that it brings forth, such as inequality and great poverty, sanctions and supports much of the destruction of the environment that exists in the world. It is not far from the truth to label economic neo-liberalism as a cruel social Darwinism or Malthusianism dressed up in the lambskin of the laudatory word "liberal."

However, for the moment, we wish to leave aside these pernicious outcomes of the individualistic creed. We wish to state a simple truth that undermines the creed. The truth is that Merleau-Ponty's findings, concerning the human body and its involvement in the field of perception, clearly indicate that all of the statements and writings of the individualistic creed are not ontologically established. They are not ontologically established because they ignore the fact that, already in establishing a field of perception, I bodily share the world with others. But ignoring ontology does not make it vanish. One conclusion is that, since individualism has no ontological foundation, it is a creed that has no philosophical ground. Probably, it would be more correct to state that individualism is established on the shifting sands of deceit.

Lacking a firm ontological foundation, the writings and scientific essays of persons who accept the creed of individualism, and many other essays and scientific theories that partially support the individualistic creed, contain, at best, partial truths. It is quite evident that these partial truths frequently reflect the whims and wishes of those persons and those political and economical organizations who wielded power when the writings were presented. Much has been written about how Darwin's theory concerning the origin of the species very well fits Victorian individualism and the flourishing of English capitalism and imperialism. We assume that Darwin's writings and scientific theory eased the conscience of many nineteenth-century capitalists. Similarly, Hobbes' essay *Leviathan* fits in very well with the emergence of capitalism in a traditional monarchy.

We shall not explore this historical-philosophical theme. We do want to say again that, in all the writings of the philosophical promoters of the individualistic creed, there is no ontological statement that relates to the fact that on the primal level of existence and perception, we bodily share the world with other beings. Consequently, the creed of individualism obscures primal truths about human existence. In addition, and on the basis of such obscuring of truth, individualism promotes the view that we human beings, as individuals, should seek power and wealth in order to dominate other beings that exist in the world. Let us repeat this point. Individualism supports the belief that human beings should dominate the world and not share it with other beings.

The results are there in the world for whoever wishes to see them. In many cases, promoters and supporters of the individualist creed, who wield power today in many Western countries and especially in the United States, have no qualms about destroying natural species and spreading ruin in the world of nature. How otherwise can we explain the widespread indifference to the ongoing persistent destruction of the natural environment that we daily witness? How otherwise can we explain this widespread indifference to such persistent destruction of natural beings and natural beauty in the United States—an indifference that has continued for the forty years since Rachel Carson published her best-selling and illuminating *Silent Spring*?

An additional outcome of the specious and ontologically unfounded thinking that promotes individualism has to do with freedom, to which we turn in the next chapter. Here we can say that, in addition to being ontologically unfounded, the writings of the supporters of the individualistic creed describe a distorted idea of freedom. Recall John Stewart Mill supporting the Opium Wars. They overlook the important fact, which has been illuminated by scores of writers from Plato to Hannah Arendt: freedom that is not shared is banal and insipid, at best. Think, as Plato did, of the tyrant. By definition, the tyrant is an individual who has seized political power, and who refuses to share his or her power and freedom with other persons in the polis. Hence, as Plato repeatedly explains, the tyrant's wielding of power is evil and will only bring forth evil deeds. Is it not true that today the individualist princes of corporate capitalism are tyrants? Dare we overlook the day-to-day widespread ruining of the environment that these individualist princes oversee and bring forth?

Finally, in contrast to the prevailing creed of individualism, Merleau-Ponty's presentation of the person's bodily involvement in the field of perception, and its sharing this field with other beings, has an ontological foundation. His ideas also provide a basis for a life in freedom that can lead to a worthy existence, in which persons respect those natural beings with whom we share the world.

NOTES

1. See Maurice Merleau-Ponty, *Phenomenology of Perception*, trans. Colin Smith (London: Routledge & Kegan Paul, 1962), 216.
2. Ibid., 328.
3. David Cummings (ed), *Encyclopedic Dictionary of Classical Music* (New York: Random House, 1997), 53–4.
4. See Merleau-Ponty, *Phenomenology of Perception*, 217.
5. See John Milton, *Paradise Lost and Other Poems* (New York: New American Library, 1961), 128.
6. See Merleau-Ponty, *Phenomenology of Perception*, 203.
7. Ibid., 206.
8. See for instance the essay: Claude Lefort, "Flesh and Otherness," in Galen A. Johnson and Michael B. Smith, *Ontology and Alterity in Merleau-Ponty* (Evanston, IL: Northwestern University Press, 1990). Responses to this essay that appear in the volume also do not address the problem of how the child establishes fields of perception.
9. William James, *Psychology (Briefer Course)* (London: Collier Books, 1962), 29.
10. For some of the major scientific flaws of neo-Darwinism see: Mae-Wan Ho, *Genetic Engineering, Dream or Nightmare: The Brave New World of Bad Science and Big Business* (Penang, Malaysia: Third World Network, 1998). Describing the scientific flaws of classical Darwinism would take us too far afield.

6

Freedom and Perception

In his endeavors to present a phenomenology of perception, Merleau-Ponty's thinking is primarily based on elucidating an ontology of perception. The goal of this ontology is to describe those moments of human existence in which a person perceives beings. His thinking, however, is in the tradition of existentialist thought that commenced with Kierkegaard. Freedom was always a central concern of the existentialist thinkers. Merleau-Ponty is no exception. In the ontology that he elucidates, he describes the being of everyday human consciousness as freedom. Yet, as already indicated, in describing human consciousness and the phenomenology of perception, and in linking this description to freedom, Merleau-Ponty provided a unique contribution.

This contribution, we have noted, stems from Merleau-Ponty's emphasis on the human body. Previous chapters have partially disclosed the manifold manners of involvement of the human body in the world—in perception, in judgment, and in action. This involvement is central to Merleau-Ponty's description of our acts of consciousness. Here we shall assume that the involvement of the human body and consciousness in perception and in the world is also central to his thinking on freedom. The assumption is based on the fact that, for Merleau-Ponty, freedom and perception blend in everyday existence.

In Chapter 5, we already hinted that Merleau-Ponty's emphasis on the body's perceptual involvement in the world, and on its links to acts of consciousness, supports the tenet that human freedom emerges in the pre-reflective consciousness. This tenet was articulately formulated and explained in detail by Jean-Paul Sartre in his philosophical tome, *Being and Nothingness*.[1] Like Sartre, Merleau-Ponty learns much about the manifestations of human freedom in everyday life by examining the workings of the pre-reflective consciousness. He accepts much of Sartre's wisdom that appears in *Being and Nothingness*. Merleau-Ponty especially relies upon Sartre's description of the relations of being-for-itself to being-in-itself. However, Merleau-Ponty, at times, goes beyond Sartre's description of the pre-reflective consciousness—for instance, when he examines the manners by which freedom emerges in our daily bodily involvement in perceiving the world and in establishing fields of perception. Put differently, Merleau-Ponty's thinking adds the dimension of bodily involvement to our comprehension of everyday perception and to human freedom. This dimension points to and elucidates additional perspectives in the everyday engagements of the pre-reflective consciousness.

The field of perception which Merleau-Ponty described and studied is a most important perspective for our understanding of the link between freedom and perception. Each person is bodily involved in establishing fields of perception and perceiving and living in these fields. Put differently, fields of perceptions are crucial to every person's pre-reflective manner of perceiving the world and relating to it. Persons express their freedom within the fields of perception that they establish.

Consider again one of the examples presented in Chapter 5. We pointed out that one realm in which Ludwig van Beethoven expressed his creative freedom was within his auditory perceptual field, and especially in the auditory field of classical music. Thanks to his freedom in that field, Beethoven composed great music. However, if the young Beethoven had never studied music, if he had not decided to become a pupil of Neefe, and later of Haydn, if he had not daily decided to study music in order to develop his musical perceptual field—he probably would not have developed that perceptual field to the breadth and depth that it reached. Perhaps he would not have become a composer.

One immediate conclusion becomes evident when we comprehend Merleau-Ponty's descriptions of perception, the field of perception, and the pre-reflective consciousness. For Merleau-Ponty, as for other existentialist thinkers, the primal emergence of human freedom is prior to reflective deliberation—because the pre-reflective consciousness is free. This freedom emerges daily in a person's bodily engagements in a perceptual field.

In order to better comprehend the linkage between freedom and perception, we look at one of Merleau-Ponty's statements concerning the field of perception, in which he emphasizes the ontological primacy of a person's field of existence. We are aware that this statement is not free of problems.

> I see a surface color because I have a visual field, and because the arrangement of the field leads my gaze to the surface—I perceive a thing because I have a field of existence and because each phenomenon, on its appearance, attracts towards that field the whole of my body as a system of perceptual powers.[2]

We admit that the citation is problematic. Hence, we purposely skip the question whether an "arrangement of the field" can "lead my gaze" to a specific surface or object. We also skip the question whether "each phenomenon" can attract the whole of my body toward its field. We are skeptical of these phrases and assertions. We can grant Merleau-Ponty that, perhaps, these problematic formulations point to specific ideas that may be worthy of exploring. These ideas, however, are beyond the scope of our study.

We do believe that, even if the wording of Merleau-Ponty's statement is problematic, his statement does disclose important truths. One truth, which we have already discussed, is that the field of perception is ontologically primary to any specific perception that is perceived in that field. In this citation, Merleau-Ponty adds two significant terms which can teach us about freedom and perception. The terms are "field of existence," and the "perceptual powers" that exist and are expressed in each person's field of existence. These terms suggest that freedom underlies a person's "perceptual powers" as they are expressed within a "field of existence." How does this conclusion emerge?

The citation suggests that every person directs his or her perceptual powers to establish fields of perception; blended together, these many fields of perception establish the person's field of existence. It is evident that every person is free to direct his or her perceptual powers, which are bodily expressions of freedom, in many directions. The directions that a person chooses to direct his or her perceptual powers establish his or her fields of perception and these, in turn, establish his or her field of existence. Thus, the choosing of the direction of one's perceptual powers involves many acts of freedom and many decisions. Most of the decisions involve a person's body and are on the pre-reflective level. Thus, a youth may direct his or her perceptual powers to playing on a basketball team or to developing one's skills in tennis. Such a decision, or series of daily decisions, will influence that youth's entire field of existence. The same is true of music, art, mathematics, botany, and many other areas, for instance, camel-riding. Developing one's skills and perceptual powers in any such area influences a person's field of existence.

Consequently, freedom underlies the establishment of each person's fields of perception and his or her unique field of existence. The fact that each field of perception and field of existence are based on freedom determines the ontological primacy of these fields in a person's existence. As a corollary we recognize that when the possibility to live a life of freedom is harshly limited, a person's perceptual powers and field of existence become confined and narrow. Here is an example that describes in some detail the relationship of freedom to a person's socioeconomic situation in the world. Frequently, the example shows, this relationship is crucial, since it determines a person's bodily life, fields of perception, and field of existence.

I delight in a musical theme that appears in the allegro moderato of Sibelius's Concerto for Violin and Orchestra. One reason for my delight is that I possess an auditory field that is acquainted with classical music and is sensitive to the performance of classical themes. The establishing of my auditory field that appreciates classical music is a result of my many decisions during my life to listen to performances and recordings of classical music. Many of my auditory decisions concerning classical music were pre-reflective, for instance, when I was attending a concert. However, I should add a crucial point. Many of my auditory decisions were possible because of my favorable social and economic situation, which allows me to live as a bourgeois.

Indeed, the example becomes vivid and enlightening if I contrast my delight in the theme of the allegro moderato of Sibelius's violin concerto to the response of a person whose situation in the world determines that he or she be totally alien to classical music. Consider, for instance, an impoverished peasant in Guatemala who can hardly earn enough to feed his or her family. Let us also assume, because it is realistic, that the peasant fears the widespread terror instigated by the thugs employed by the ruling classes in Guatemala. This everyday bodily fear is involved in and frequently determines the peasant's daily choices. Thus, because of fear, the peasant does not join a peasant's union or attempt to openly struggle for a better standard of living. According to Kofi Anan, the Secretary General of the United Nations, one billion two hundred million inhabitants of the earth live in abject poverty, which means that they live on less than one dollar a day. We add that, in many areas of the world, these poor people are peasants, who also live in great fear of the violence of the thugs employed by the ruling classes. Hence, the impoverished terrorized peasant from Guatemala is hardly exceptional in today's world.[3]

It is almost certain that for a Guatemalan peasant who lives in abject poverty, as for almost all the hundreds of millions of impoverished of the world, the auditory field of classical music does not exist. Hence, the musical theme from the allegro moderato of Sibelius's violin concerto may sound to the peasant like a mere sequence of sounds. We are aware that many bourgeois and affluent people also live with impoverished fields of perception. Some of them might relate like an impoverished peasant from Guatemala to Sibelius's violin concerto. We will say more about such choices in the following chapters.

We can broaden this example, and safely assume that the field of existence of the peasant who lives in debilitating poverty is greatly impoverished by his or her lack of many additional worthy fields of perception, say, of the arts and sciences, of medical aid, and of basic economics. We can also assume that the perceptual powers of the peasant are almost always directed to areas that are concerned with personal bodily survival, and that this emphasis on survival impoverishes his or her field of existence.

A major social and existential conclusion emerges from the example and from Merleau-Ponty's concepts and thinking. Consider a person growing up in affluent circumstances—we shall soon mention a literary example of such a person. Such a person has the opportunity to establish many fields of perception, and these fields of perception may include many dimensions; hence, the person can choose to live his or her freedom in many fields. In contrast, life in abject poverty and in fear of terror organized by the ruling class—as experienced by the impoverished Guatemalan peasant and many others in similar circumstances—confines a person's freedom and perceptual fields.

We can elucidate this conclusion with the help of Merleau-Ponty's terms. A harshly oppressed life in which one struggles for bodily survival in a milieu of scarcity—a life such as that lived by the Guatemalan peasant—reduces and confines the impoverished person's fields of perception, field of existence, and perceptual powers. In brief, the impoverished person's freedom is reduced and confined, as are his or her field of existence and perceptual powers. Consequently, Merleau-Ponty's ontology reveals that the richness or the poverty of a person's field of existence is frequently a result of the opportunity that person has to live one's freedom beyond the daily struggle for mere survival.

Since this point is crucial for understanding much of the ruining of the environment, we want to mention two conclusions and present an additional literary example. There is a dynamic relationship, bordering on the causal, between the possibility to establish fields of perception and the possibility of fully living one's freedom. A person's economic and social circumstances will very often dictate the possibilities of his or her bodily existence, and with these possibilities, his or her opportunities of establishing specific fields of perception and living in freedom.

Consider again the life of Marcel, the bourgeois narrator in Proust's *Remembrance of Things Past*. Marcel's family was very affluent, and he was free from the economic struggle for bodily survival. Hence he was born in a situation that allowed him to develop many exquisite fields of perception, in which he was bodily active. These fields of perception included, in his youth, that of blossoming hawthorns and of shimmering water lilies floating in a picturesque river; later, Marcel developed fields of perception which included the seeing of the beauty of a painting, a work of architecture, and a musical compositions.[4]

Evidently, harsh economic circumstances, such as living a life of scarcity, greatly reduce the possibilities and opportunities for developing fields of perception such as those developed by Proust's Marcel. Yes, even in Proust's novel, from the lives of other characters, we learn that a life of scarcity limits the person's fields of perception and perceptual powers. Poverty confines and impoverishes the poor person's body and field of existence.

Someone may ask: How is this ontological discussion of freedom and its relationship to fields of existence and fields of perception linked to relating to the environment?

One immediate answer emerges from the brief discussion of the peasant from Guatemala. A peasant who is daily exploited by wealthy and cruel landowners and who lives in abject poverty—whose freedom is primarily engaged in working many hours in order to obtain the necessities of life—will primarily direct his or her perceptual powers in the direction of bodily survival. Consequently, the fields of perception of any person who exists in abject poverty are usually limited and confined. They rarely include the fields of perception which can illuminate the political and economic decisions that bring about his or her ongoing exploitation. Indeed, the peasant's fields of perception will rarely transcend his or her daily attempts to cope with the harsh economic circumstances which confront him or her. Consequently, for the impoverished downtrodden person, the initial perception of beings in the world, and one's confined fields of perception, frequently exclude the possibility of sharing one's being with other beings in the world—say, with lizards or spiders or lions. Such a sharing will rarely enter the impoverished downtrodden person's field of existence, except, perhaps, through social or religious rites.

Merleau-Ponty explains in detail that for an impoverished exploited person to recognize that he or she is exploited, that person must change one's field of perception. The new field of perception emphasizes the fact that one's plight—being exploited, being poor—is shared by many other people, but not by all. Some people are wealthy and participate directly, or in a roundabout manner, in the exploiting of those who are impoverished. With such a new field of perception, the exploited person may choose to act on the fact that his or her harsh unjust situation is shared by many other human beings in one's immediate circumstances, and in the world. Such action frequently brings with it new fields of perception.[5] We believe that only after such a change in one's fields of perception will the exploited impoverished person be open to the possibility of relating to other natural beings and species that exist in the world.

We acknowledge that among those people who have lived in poverty, there have been exceptions. One such exception was Mohandas Karamchand Gandhi (Mahatma Gandhi), who chose to live in poverty and taught persons to generously share their being with other beings that exist in the world. But such exceptional spiritual leaders are rare. As a rule, and especially after the triumph of capitalism in most areas of the world, persons who are born into abject poverty are trapped in that poverty. In many respects, these human beings are bodily trapped in an economic maze—with no exit. These needy and trapped people are concerned with the daily survival of their body, not with sharing their being with other beings.

This immediate sad answer to the above question is not our final statement. There exists a more encouraging answer that includes broader outlook on the relationship of freedom to sharing the world with other beings.

Let us assume that we human beings wish, in our life on earth, to share the world with other natural beings. We have pointed out that elements of an ontology that accords with this worthy wish emerge in Merleau-Ponty's description of human perception. Understanding the implications of these ontological findings can, at least partially, help us to grasp that sharing the world with other beings is central to perception and to human existence. My field of existence is based on my bodily shared fields of perception. Thus, in order to try and to bring our wish to share the world with other beings to realization, we can attempt to learn to live in the world in a manner that relies upon our original perceptual relation with beings in the world. Such an attempt to learn to share one's being with others is an act of freedom, and a challenge.

In order to undertake such a challenge, while learning from Merleau-Ponty's thinking, we must address the problem of human freedom—in all its manifestations. In his short chapter on freedom in the final pages of *Phenomenology of Perception,* Merleau-Ponty gives some examples of the links between perception and freedom. But these examples are firmly linked to his Marxist views on the social and economic relations that prevail in capitalist countries. Furthermore, the examples seem to be founded upon the optimistic belief that it is possible to develop class-consciousness among the laborers, and that this class-consciousness may even lead to a freedom-enhancing proletarian revolution. Political developments in the so-called communist countries, that existed at the time Merleau-Ponty wrote his study, have greatly altered our understanding of communism as it was lived in the twentieth century in the Soviet Union and its satellite countries. These political developments occurred after Merleau-Ponty's death, and have cast grave doubts upon the validity and benefits of a proletarian revolution as he seems to have perceived it. Hence, we shall not cite Merleau-Ponty's examples.[6] Instead we shall attempt to relate to what we believe to be a more foundational, and yet factual, level of this linkage between freedom and perception.

Look again at our example of Sibelius's violin concerto and the existential plight of the peasant from Guatemala. We already stated some of the conclusions of the example. A person whose freedom is confined to the struggle for bodily survival—such as the billion and two hundred million people who today live in abject poverty—will only rarely be able to direct his or her perceptual powers to transcend bodily needs. That person's fields of perception and field of existence will be very limited. Usually, such a person will not wish or find the opportunities to share his or her being with other beings. Now we can add that most of these people are impoverished because of the corporate capitalistic economic regime that currently rules much of the world.

Consequently, from a slightly different perspective, we return to a major fact presented in previous chapters. We stated that the currently prevailing political and economic regime, which emphasizes corporate capitalism and promotes globalization with all its evil manifestations, stifles the possibility of sharing the earth with other beings. We now add to this statement the sad fact that this regime condemns one fifth of the human beings living on the face of the earth to existing in abject poverty. It traps these human beings in this debilitating maze of poverty. Hence, the current world-dominating political and economic capitalist regime pretty much eliminates the possibility that these impoverished people will struggle to realize an ontology of sharing the world with other beings, and will wish to share their being with other natural species.

We believe that many of the princes of corporate capitalism throughout the world know the sad outcomes of their pernicious policies. We believe that they comprehend that, usually, the being of the abjectly impoverished person is incarcerated within a few limited and meager fields of perception. Hence, they know that the impoverished person will only rarely attempt to establish a field of existence that includes sharing one's being with other beings in the world. We venture to suggest that, when confronted with these sad results of their own political and economic policies, choices, and actions, these princes of corporate capitalism couldn't care less.

We can now admit that the earlier question as to the links between freedom and the ontology of perception is central to the theme of this book. Hence, we wish to further elucidate the relationship between freedom, a field of perception, a field of existence, and sharing the environment with other species. In order to do so, we must address one of the iniquitous developments of the twentieth century, a development that Merleau-Ponty did not discuss at length. We refer to the fact that people have been encouraged by government agencies, and by the mass media that usually serves the government, to develop perceptual fields that are constructed upon deceitful ideas and false facts.

During the twentieth century, probably the most evil example of such a perceptual field constructed on deceitful ideas was developed by Hitler and his Nazi ideologues and propagandists. (The perceptual field promoted by Stalin and his cohorts comes in a close second.) According to this Nazi field of perception, embraced by many simple Germans, all Jews and

Gypsies, and some other ethnic and religious groups and human communities, were considered to be parasites and pests that should be exterminated. Fortunately, the Nazis were defeated in a world war, and the fields of perception that they vehemently promoted were broadly condemned by the victors and by the world community. But, unfortunately, many of the methods for spreading fields of perception that are based on deceit have been adopted by other, less evil, regimes. You may frequently encounter these methods in countries that are called democratic. Consider, in brief, a current example.

Today, President George W. Bush and his host of pernicious advisors and sycophants are frequently attempting to establish fields of perception that are based on deceit. For instance, these leading Americans use the word "terror" to attempt to establish a field of perception that benefits their interests as rulers. They employ the word "terror" for this purpose both within the United States and in the world. The incident of "terror" that is used to establish this field of perception is the September 11, 2001, attacks on the World Trade Center and the Pentagon in the United States. Hence, almost by definition, "terror" consists of attacks on the United States and on its allies, say, on Israel or Colombia.

According to this distinction, the brutal oppression of the Palestinians in the Occupied Territories by Israeli armed forces, or of the Colombian peasants by right-wing paramilitary organizations, which are supported by the Colombian army and the United States, is never termed "terror." When Israelis are killed by Palestinian suicide bombers, they are victims of terror. When Colombian capitalists are killed by people whom they harshly oppressed, these capitalists have been murdered by terrorists. However, when Palestinian children are killed by Israeli soldiers (at least fifty Palestinian boys and girls under the age of eight have been killed in the past two years) or when Palestinian pregnant women die in childbirth because Israeli soldiers do not allow them to reach a hospital, or when dozens of Colombian peasants are massacred at the hands of the right-wing paramilitary organizations—in all these cases, those who have died are never, but never, victims of "terror."

It is very sad that the propagators of this perceptual field of "terrorism" in the United States dominate both the mainstream and right-wing media. With the help of the media, these propagators use the field of perception characterized by "terror" in order to justify the evil deeds and the rhetoric of their government. Prominent among the government evils of the Bush administration is the fact that the United States government has decided to defy constitutional principles within the United States. In addition, the Bush administration has decided to use its economic powers and military force wherever it wishes in the world, without caring for the widespread ruin and the human suffering that result from its deeds. Here is our major point. All of these evil deeds are described in favorable terms in the fields of perception, that are based on deceit, and that the United States government and the compliant media establish and spread.

We shall skip additional historical examples of how evil politicians establish and spread fields of perception that are based on mendacious claims. They abound in our daily existence, and are not difficult to perceive. However, we do want to respond to a question that arises. The question is: Given this spread of fields of perception that are based on deceit—a phenomenon which Merleau-Ponty did not discuss at length—how can a person cope with these deceitful fields of perception? Or, put differently, how can a person struggle against accepting these evil mendacious fields of perception?

The best answer that we can suggest is to return to the foundation of ancient philosophical wisdom, especially as it emerged in ancient Greece. That wisdom was based on thinking. Thinking came into being in ancient Greece together with the belief in the significance of questioning. Socrates and his fellow thinkers believed that we must question every proffered idea, every perception, every statement, every historical event. Learning from these thinkers, we hold that persons must strive to direct their perceptual powers to a field of existence in which questioning is paramount and in which the search for truth is prominent. Hopefully, living in such a field of existence will help those people who decide to seek truth and to question, see, and reject the deceit that exists in many currently proffered fields of perception.

We do want to present a historical example of one extreme perceptual field, that was constructed upon mendacious claims and relates to the theme of this book. This blatantly mendacious field of perception was established in the United States in the first six decades of the twentieth century—during the two world wars, in the period between them, and in the years following World War II. This field of perception was widely publicized and has not yet vanished from contemporary life in the United States. The major promoters of this mendacious field of perception were companies and corporations that dominated the United States chemical industry and many of the supporters of chemical warfare within the government and the army. These promoters of this field of perception were backed by politicians and by some prominent scientists.

Working together, these groups of scientists, politicians, government bureaucrats, and chemical company executives endeavored to spread the belief that there is an immediate need to use chemical compounds and poisons so as to eradicate the entire class of insects. They argued that all the insects that exist in the world are enemies of the human race and of human existence.—No less!

In his book, *War and Nature: Fighting Humans and Insects from World War I to "Silent Spring,"* Edmund Russell gives many examples of the attempts made in the United States to convince the public that all insects are enemies of humankind. For instance, in 1921,

L. O. Howard, chief of the United States Bureau of Entomology, chose in his address as retiring president of the association to promote a new image for entomology. His speech framed "the war of humanity against the class *Insecta*" as "the next great world war."[7]

Note that Howard was speaking as a scientist. His rhetoric was presented so as to arouse fears and to promote the idea that human beings must daily fight against all insects. Let us frame his deed using Merleau-Ponty's terms. The chief entomologist of the United States, L. O. Howard, endeavored to establish a field of perception in which all species of insects and all insects—the entire class *Insecta*—are enemies of the human race.

Russell points out that Howard's ideas were expanded upon in the media and were hardly challenged by the scientific establishment. Scientific establishments, as has been repeatedly shown even in countries that boast a democratic regime, are not always wedded to truth and to spreading truths. Quite often many scientists are lackeys of a company or of a regime.

Russell also provides many examples in which Howard's ideas were promoted by popular writers. Here are Russell's citations from an article by William Atherton Du Puy titled "The Insects Are Winning," which appeared in *Harper's Magazine* in 1925. Russell explains that Du Puy's article abounds in rhetoric against the "insect menace" that has waged a "thousand year war against the human race." He specifically cites the following:

The issue is vital: no less than the life or death of the human race. If man wins he will be the dominant species on this earth. If he loses he will be wiped out by this, his most ambitious racial enemy.[8]

We shall not attempt here to address the gross stupidity dripping from the above two citations, that is, from the writings of Howard and Du Puy. (We do address this stupidity briefly in Chapter 7.) Nor do we wish to attempt to prove what is evident: the views have no scientific or factual foundation. It has been known for many decades that the billions of insects that exist in the world, in addition to being a spectacle that may arouse human wonder and awe, are necessary for the continual survival of the human race, and of other species in the world—many species of flowers, for instance.

What is significant is that, despite the stupidity and deceit of their pronouncements, the promoters of a world war against the class of insects pretty much succeeded in establishing the field of perception that supported their aims. They aroused widespread fears that changed many people's perception of insects and of nature. Learning from Merleau-Ponty, we can state that they constructed a field of perception, based on deceit, in which chemical compounds and poisons can assist us in fighting insects and saving humanity. According to these writers, the chemical companies will save us from the class of insects—who constitute humanity's "most ambitious racial enemy." Thanks to the broad acceptance and spread of this field of perception in the United States, the chemical companies sold millions of tons of chemicals that destroyed insects. Now we know that these poisons also brought great destruction to the environment and to human and animal life on earth.

Edmund Russell points out that the accepted war against insects, and the perception that insects constitute enemies of humankind, received additional impetus during World War II, and especially in the war of the United States Army in the Pacific against the Japanese. One reason for the revived impetus to eradicate insects was that it became known that, in the first two years of the war in the Pacific, for every American soldier killed in battle, ten soldiers were killed by malaria. As a result, the anopheles mosquito, which spread the malaria parasite, became recognized as an enemy of the American people.[9] We want to add an evident point. The justified war against malaria, which continues today, should not cloud our perceptions as to our relations with all other insects and with nature.

To recapitulate, Russell's book, *War and Nature,* reveals that since World War I, for about half a century, the erroneous field of perception that viewed insects as enemies of the human race was promoted by chemical companies and scientists. This mendacious field became widespread in the United States, and probably in many other countries. He notes that advertising, humor, popular writing, and so-called scientific scholarship all contributed to legitimizing and spreading this erroneous field of perception. We would add that the widespread deceit that was instigated by corporate capitalist chemical companies who were greedy for profits—Russell names some of these companies—underlay the wish to establish this erroneous field of perception.

In the past four decades many people have questioned the validity of this erroneous field of perception, and have shown its terrible mistakes. These mistakes, many of which have been stated by scientists, are not dealt with in Russell's book. Thanks to these doubts, laypeople's acceptance of the field of perception that vilified insects began to diminish. However, as Russell shows, the field of perception that sanctioned total war against insects slowly began to change in the United States only in the 1960s. Probably a major influence in bringing about this change was Rachel Carson's widely read study, *Silent Spring,* published in 1962.[10]

Today, we do not believe that we should be overly optimistic. The question still arises: Today, more than forty years after *Silent Spring,* has the stupid field of perception that condemns all insects to extinction, a field of perception promoted by Howard and other Americans, changed? We have no firm evidence for such a change, hence we would answer: Probably a bit, but we are unable to point to the exact direction of change. We do know that large chemical companies in the United States, and in other developed countries, continue to produce and sell millions of tons of poisons that are used against insects.

In summary, Merleau-Ponty teaches that human freedom emerges in the perceptual field, and also that freedom bodily establishes the perceptual field. Together, the perceptual fields that a person develops establish his or her field of existence. These statements are not circular. Think again of Beethoven. His freedom emerged in the auditory field of perception that was classical music—he composed and published three piano sonatas at the age of thirteen. However, in order to be able to compose music, Beethoven chose to establish and to personally develop an auditory musical field of perception. And, it was on the basis of, and within, that personal field of perception that Beethoven could freely compose works of original music.

Freedom, we have also suggested, allows each person to courageously question the perceptual fields, the ideas, and the world that he or she encounter. Merleau-Ponty stresses that the oppressed day-laborer in France can always decide to courageously question the prevailing field of perception that allows the capitalist to daily exploit him or her and many other laborers. Such courageous questioning may lead the laborer to a realization of his or her freedom and, perhaps, to a recognition of the need to personally struggle for a world in which that freedom will be respected. Consequently, when courageous questioning is authentic, it is frequently an assuming of responsibility for one's life and for the world that one perceives and in which one exists.

It is evident that, today, such courageous authentic questioning is very rare. Such is unfortunate. Authentic questioning is especially needed when we contemplate the many mendacious fields of perception that we are daily encouraged to adopt by politicians, advertisers, and the media. Furthermore, as we have shown in this book, much radical questioning is sorely needed if we wish to halt the rampant destruction of natural species, and of the planet as we know it.

NOTES

1. Jean-Paul Sartre, *Being and Nothingness*, trans. Hazel E. Barnes (New York: Washington Square Press, 1956).

2. See Maurice Merleau-Ponty, *Phenomenology of Perception*, trans. Colin Smith (London: Routledge & Kegan Paul, 1962), 318.

3. For more on the plight of peasants in Guatemala see: *I Rigoberta Menchu: An Indian Woman in Guatemala*, ed. Elisabeth Burgos-Debray, trans. Ann Wright (London: Verso, 1984).

4. Marcel Proust, *Remembrance of Things Past,* trans. C.K. Scott Moncrieff (New York: Random House, 1934).

5. See Merleau-Ponty, *Phenomenology of Perception,* 444.

6. Ibid.

7. Edmund Russell, *War and Nature: Fighting Humans and Insects with Chemicals from World War I to "Silent Spring"* (Cambridge: Cambridge University Press, 2001), 75–6.

8. Ibid., 77.

9. Ibid., 95–118.

10. Rachel Carson, *Silent Spring* (London: Penguin Books, 1965).

7

Responsibility

We have pointed out that, according to many existentialist thinkers, freedom is the ontological foundation of all human endeavors. Such an understanding of human existence and reality means that a person is responsible for all the choices that he or she makes in daily life. Many of these choices, existentialists pointed out, may seem to be minor or irrelevant; however, they are not insignificant. Even minor choices, they hold, assert, determine, and reflect a person's manner of Being-in-the-world. Put otherwise, many minor daily choices participate in determining a person's character and life.

Consider Kierkegaard's "Diary of the Seducer."[1] In this illuminating narrative, Kierkegaard shows that Johannes's many minor, seemingly irrelevant choices contribute substantially to his project of seducing Cordelia. Indeed, we learn from Kierkegaard how Johannes's many minor choices throughout the day, or the month, or the year, determine Johannes's becoming and being a seducer. Much the same is true of a personal decision to live solely for one's aesthetic pleasures, as Johannes did. Kierkegaard perceived many persons in his time who lived solely for aesthetic pleasures, as do many persons living today. Such a choice of a way of life also emerges in many minor choices that the person makes every single day. Think now of the decision to strive to assume responsibility for other persons and for other beings that exist in the world. Many minor and major choices determine whether, when, how, and where such responsibility emerges in a person's life.

Merleau-Ponty's writings are considered to be in the tradition of existentialist and phenomenological thinking. Yet, in *Phenomenology of Perception* he did not discuss responsibility at any length. We mentioned that the final chapter of his book is dedicated to freedom. But, in this chapter, not very much is said about our personal responsibility for other persons or other beings that exist in the world. Probably the closest that Merleau-Ponty gets to asserting human responsibility for others is the following statement from the last page of his book: "*your* freedom cannot be willed without leaving behind its singular relevance, and without willing freedom *for all*."[2] But this statement is abstract, and does not relate directly to his findings concerning perception.

Despite its not dealing directly with responsibility, we believe that much can be learned about assuming responsibility for other beings from some of the findings that are described in *Phenomenology of Perception*. We especially learn from the facts that Merleau-Ponty presented concerning each person's fields of perception. From what we have shown in previous chapters it is evident that I am bodily involved in any field of perception that I perceive and establish. This bodily involvement already includes at least partial responsibility for establishing this particular field of perception and for my interactions with the beings that are included in the field. To mention an extreme example, the Ku Klux Klan racist is responsible for establishing a field of perception in which he or she perceives all Jews or "niggers" as subhuman. Merleau-Ponty also pointed out that my intentions are central to establishing a specific field of perception. In what follows, we shall suggest in what additional manners these findings may have implications concerning our responsibility for other persons and for other beings. Many of the implications concerning responsibility that we present go beyond what Merleau-Ponty stated in his study of perception.

However, before turning to the links between responsibility for other beings and species that are found in the world and the fields of perception that a person establishes in his life, a poignant question should be posed. The question is: Why should a person decide to assume responsibility toward other beings and other species in the world, and what should be the scope of his or her responsibility? We want to present our partial answer to this question. We are not sure that Merleau-Ponty ever addressed this question straightforwardly. In our answer, we primarily indicate a worthy direction for human existence. We do not present a catechism.

Our abbreviated answer to the question in the preceding paragraph commences with a simple belief. We believe that every person, every human being is sojourner upon the face of the earth. What is more, unlike all other beings and species that are found upon the earth, every human being knows that he or she is a sojourner. Indeed, every adult human being knows that the time that he or she will live is limited. Every human being knows that one day he or she will die. Each human being also assumes that, after one's death, many other people and other generations of human beings will continue to reside upon the earth.

This knowledge already includes a question: How should I live, knowing that I am merely a sojourner upon earth? Our answer learns from the wisdom of those outstanding persons who endeavored to enhance the spiritual and personal existence of their fellow sojourners upon earth—examples of such persons are the Hebrew prophets, Sophocles, Plato, and in the twentieth century, Anna Akhmatova and Mohandas Gandhi. The answer emerging from the words and deeds of these leaders of human spirituality is that I should live a life in which I pursue truth, justice, and wisdom. Of course, there are other worthy things to pursue, for instance, love, beauty, and friendship. But, for most of the spiritual leaders, truth, justice, and wisdom are prior. We should add that many of these spiritual leaders believed that the pursuit of justice requires living in a manner that does not rob the generations that will inhabit the earth after one's death of their full life, and of the richness of nature that currently exists upon the earth.

Make no mistake, however. The spiritual leaders from whom we can learn to live a worthy life knew that undertaking a challenge to pursue truth, justice, and wisdom is very difficult. Hence, they stated that, at least, a person should not spread deceit and ruin—neither in one's society, nor in the world.

The important point for the theme of this book is that a worthy human existence must include the pursuit of truth, justice, and wisdom. In addition, at the very least a person should strive not to spread deceit and ruin in the world. Thus, if I wish to promote a worthy human existence upon earth, including in the years after my death, I must do my utmost to halt the spread of deceit and ruin that I encounter, and I must encourage the pursuit of truth, justice, and wisdom. Such a life, the spiritual leaders suggested, is a life of wholeness and integrity.

We believe and hold that, as a human being who can think and act, promoting such a worthy existence is my responsibility toward other persons, including those persons that will be born after my death. My responsibility to help preserve other natural beings that exist in the world is included in the principles of such a worthy existence. Thus, the scope of my responsibility cannot be precisely determined. We can say that it includes, in addition to the pursuit of truth, justice, wisdom, and other worthy things, a responsibility toward the environment and the many beings of nature. Put otherwise, while pursuing those things that are worthy in themselves such as truth, justice, and wisdom, I should struggle daily to preserve other species and beings in their natural environment. After all, this natural environment is the world in which we human beings exist, and in which human history developed and brought forth both spiritual leaders and simple people.

Learning from the findings of *Phenomenology of Perception,* we have indicated that a person's bodily involvement in a perceptual field can constitute a first step toward assuming responsibility for the beings in that field. Here is a simple example that fits with our discussions of the field of classical music in previous chapters.

A good conductor of a symphony orchestra, for instance Riccardo Muti, is bodily involved in the field of classical music. We are confident that Muti recognizes that it is his everyday responsibility to perceive in detail the beings in this field of perception that he has established. We assume that Muti knows that, within the field of classical music, and particularly within the field that he establishes while conducting classical music, he must be sensitive toward, at least, three major participants in the field: the musical compositions that he is conducting, the musical expression that can be brought forth from the various instruments in the orchestra, and the members of the orchestra. Furthermore, he must perceive the various and complex interactions that may come into being between these different beings that exist in his perceptual field of classical music.

As a conductor, Muti has also assumed the responsibility, with the help of the members of the orchestra that he is conducting, to bring forth for audiences the beauty of certain musical compositions that are found in the field of classical music—say, Stravinsky's ballet music, *Petrushka,* or Verdi's opera, *Attila.* Muti recognizes that, as a conductor, he must lead the members of the orchestra and the soloists to perform so that each brings forth the best of his or her abilities. When such occurs, each great musical composition is performed in a manner that brings forth its complexity, uniqueness, and resounding beauty.

We have purposely chosen the field of classical music and the example of a famous conductor, Riccardo Muti. We believe that what occurs in this field can partially illuminate the responsibility that all persons have in many of the fields of perception which they establish and in which they are bodily involved. It is not difficult to comprehend that, for Muti to fulfill his responsibility in the field of classical music, he must develop his perceptual sensitivities to the many beings that exist in the field. Furthermore, listeners to classical music—and many other persons—know that the field of great classical music is, generally, not a realm in which persons endeavor to conquer, dominate, exploit, or oppress the beings that can be found in that field. Persons who are bodily involved in perceiving and acting in the field of perception of classical music usually act responsibly toward the beings in the field. Indeed, many persons strive to relate harmoniously to the beings in this field.

This example leads us to two key questions that are relevant to relating to natural beings that we encounter. Is the field of classical music very different from the other fields of perception in which human beings are bodily involved? And if some fields of perception differ from the field of classical music, in what ways do they differ?

We will not fully respond to these questions. What is more, we respond in a roundabout manner. Yet, the limited response that we present is instructive for the theme of this book.

Our response commences with a careful look at the opening lines of Vandana Shiva's book, *Stolen Harvest: The Hijacking of the Global Food Supply*. Shiva writes:

Over the past two decades every issue I have been engaged in as an ecological activist and organic intellectual has revealed that what the industrial economy calls "growth" is really a form of theft from nature and people.

It is true that cutting down forests or converting natural forests into monocultures of pine and eucalyptus for industrial raw material generates revenues and growth. But this growth is based on robbing the forest of its biodiversity and its capacity to conserve soil and water. This growth is based on robbing forest communities of their sources of food, fodder, fuel, fiber, medicine, and security from floods and drought.[3]

However a reader may respond to Shiva's assertive statements and to her condemnations of the leaders of what she calls the "industrial economy," one truth emerges. When Shiva perceives and opens a specific field of perception that includes many natural species and beings—say, a forest in India, where she resides—she has much in common with a good conductor of an orchestra. Like the conductor, Shiva has no wish to conquer, dominate, exploit, or oppress the beings that are found in the field of perception that she has opened. She has no wish to rob the forest of its biodiversity, or rob the forest community of its sustenance and way of life. She is determined *not* to ruthlessly exploit the natural beings or the human beings that she encounters in her field of perception.

Underlying this approach, the reader of her books discerns that Shiva also recognizes, like the conductor, that she must develop sensitivities to all the beings that exist in her field of perception, and to the possibility of harmony between them. This sensitivity is sorely needed if she wishes, at least partially, to perceive the complexity that arises from the possible interactions of the various beings in this field. Such sensitivities may help Shiva, as an ecological activist, to perceive and bring forth the best of each of these beings. Shiva's manner of relating also discloses truths about the beings in the forest which a conquering exploiting approach may often conceal. It is evident that by establishing such a field of perception, which repeatedly appears in her books, and which the cited paragraphs intimate that she has established, Vandana Shiva is assuming responsibility for other persons and other beings with whom she shares the world. She is also educating her readers to change their erroneous fields of perception—which are frequently utilitarian and flawed.

The comparing of Shiva's field of perception to that of a conductor of classical music leads to an additional truth found in the statements that open her book. The truth is: Vandana Shiva has established a field of perception that differs substantially from the widespread means-end fields of perception that currently prevail in our materialist-oriented world. The wholesale selling of means-end fields of perception constitutes the daily bread of exploiting capitalists and conquerors and their cohorts of supporters. Hence, for centuries, these fields of perception have been promoted, publicized, and justified by the bourgeois and by their spokespersons. Note that Shiva indicates that many of the establishers of such a means-end macroeconomic field of perception are robbers.

Yet, someone may ask: What exactly is the difference between Shiva's field of perception and the means-end fields of perception that have been established by exploiting capitalists and conquerors, and by many of their supporters and sycophants? The answer to this important question is complex, and emerges slowly in the course of this chapter.

The means-end fields of perception frequently consist of an attempt to promote and to justify the capitalist's intention to exploit human beings and natural beings for one's own financial benefit. Shiva's field of perception includes no such intention. Unfortunately, in the twenty-first century, many of the exploiting corporate capitalist fields of perception are broadly promoted and justified in the media. (In the Introduction, we mentioned the extolling of the cult of progress and the admiration of technology as manners of justifying exploitation and oppression.) Those who present and justify these means-end fields of perception include politicians, so-called philosophers, myopic scientists, well-paid journalists, and a whole range of academics who are employed by wealthy corporations. All of these justifiers of the means-end fields of perception do not relate to Shiva's arguments or to the facts that she brings; they blatantly ignore them. In the world dominated by capitalist greed, Vandana Shiva has many opponents.

The fact that Shiva's field of perception differs from that of the corporate capitalist has a few important implications that relate to the scientific endeavor as it is now grasped by many scientists. Note that the accepted paradigms of large areas of scientific endeavor are often rooted in many different means-end fields of perception. In passing, we mention that science can also include fields of observations that are not necessarily linked to a means-end field of perception. Think of botany, anthropology, or geography.

Shiva's field of perception differs from those scientists who embrace and stake their claims on the means-end field of perception. In establishing her field of perception, Shiva recognizes that the accepted and prevailing corporate capitalist fields of perception which, in a roundabout manner justify the ruining of the environment, have grave implications for human existence upon earth. Her fields of perception allow her to condemn those capitalist-serving scientists and their fields of perception. Many prominent scientists, among them E. O. Wilson from Harvard University, whom we mentioned in previous chapters, refuse to judge corporate capitalism. And whoever refuses to judge, will never condemn.

Thus, Wilson, and probably most of his illustrious academic colleagues, will never announce that capitalist methods are robbery—as Shiva did. These scientists will never describe the evils suffered by the simple people who live in a forest, and who are robbed of their livelihood and way of life by the destructive actions initiated by corporate capitalists. At times it seems that simple people are purposely excluded from these illustrious scientists' fields of perception. Hence, as a scientist, Shiva differs substantially in her views, in her fields of perception, and in her scientific findings from Wilson and from the scientists who, straightforwardly or surreptitiously, serve the greedy capitalist corporations.

Consider the scientists employed by the giant corporate agricultural-supplies company, Monsanto.[4] Monsanto is but one among many capitalist corporations that environmentalists claim continually ruin the environment. Monsanto employs many scientists who seem to care not at all about their company's goals. Quite often, these so-called scientists frequently refuse to acknowledge some well-known scientifically proved truths. For instance, they will deny facts that have been proven by environmentalists, that the capitalist corporation that employs them engages in practices that can destroy nature as we know it—for centuries to come. In addition, these scientists refuse to question the rationale that underlies much of their research and work.

But science is committed to the pursuit of truth. How do these scientists establish or justify their refusal to question the truths and the rationale that underly their research?

Some of these scientists choose to be lackeys and probably establish a field of perception in which the natural environment, and its dynamics, is totally irrelevant to human existence in the world. In that field of perception, there also is no need to disclose worthy truths or to assume responsibility for the use of the scientific findings that one has discovered. Put differently, probably most of these lackey-scientists develop a field of perception centered upon the means-end approach. Such a field of perception supports their isolated detached scientific engagements and work. The scientists hold that they have a topic to research and all accepted means of research are justified. The result of their research is, at times, reported to their scientific milieu; it is always reported to the company that employs them. Leaders in the company decide what to do with the results in order to enhance profits. This decision, many a scientist holds, is not related to his or her field of existence. Thus, the scientist's research and work, perhaps his or her entire existence, is a means for the company to gain profits.

It is not difficult to discern that adopting so-called detached fields of scientific perception, that do not relate to the results of one's endeavors, allows these scientists to be irresponsible and to act immorally. We would add that many of these scientists relate to other beings in the world as if their scientific engagement allows them to be conquerors, exploiters, and oppressors.

In contrast, Shiva's field of perception, which repeatedly emerges in her books, assumes responsibility for the environment and for human beings who exist in a specific environment.[5] As mentioned, she assumes such responsibility much as the conductor assumes responsibility for his orchestra. This responsible approach underlies Shiva's willingness to call evil people and exploiting companies robbers. It is evident that by assuming responsibility for natural beings, her field of perception differs substantially from the field of perception established by the company-scientists, say, of Monsanto, whom we briefly described. We repeat our main conclusion. Unlike Shiva, these company-scientists, and the corporations that they diligently serve, refuse to assume responsibility for other persons and for other beings that exist in the world.

We have already hinted at a second point concerning perception that emerges from Shiva's writings. The field of perception that Shiva established has much more access to truth than the field of perception promoted by the capitalists and the scientists who embrace the means-end approach. The reason becomes evident when we examine the intentionality that blends with the perception when establishing the field. As mentioned, Shiva establishes a field of perception in which sharing and sensitivity to other beings is prominent. Shiva willingly shares her body, her being, with other beings that are found in the field of perception. In contrast, the underlying intentionality that blends with the capitalist's act of perceiving is naked utilitarianism. It is safe to assume that the field of perception that a person establishes on the basis of utilitarian intentionality is skewed toward the belief that the means-end approach is the key to building relations with all beings. In other words, frequently the utilitarian superciliously relates to other beings solely in terms of their utility for oneself.

It has been repeatedly shown by Kierkegaard, Nietzsche, and many other thinkers that the means-end approach is superficial and discloses very few truths about human existence in the world. Put succinctly, many prominent thinkers have repeatedly indicated that utilitarianism has nothing to teach us about human existence and the existence of the other beings that we encounter. For instance, the process and the daily decisions by which Kierkegaard's Johannes became a seducer have almost nothing to do with utilitarianism. For one simple reason: Utilitarianism and the means-end approach deal solely with external relations. In contrast, Kierkegaard patiently describes the minor conscious decisions that constitute the development of a whole person, of Johannes, the seducer. These minor conscious decisions usually have little to do with means-end external relations.

Nietzsche stated that the means-end approach to questions of human existence is wrong, because it confines human freedom to a narrow realm in which all that matters is obtaining a specific external end. Two broadly sought external ends are money and fame. In contrast, as the ancient Greeks already knew, glory and nobility transcend the means-end approach, as do courage and wisdom, love and friendship, and many other aspects of human existence. What concerns us here is that the means-end field of perception discloses very few truths about the human beings that are found in that field. Furthermore, we can learn from Kierkegaard, Nietzsche, Buber, and other philosophers that a utilitarian field of perception is hollow and vacuous; hence, it is foolish to rely upon this field of perception in any discussion of human existence.

Consider a few brief examples. Due to its emphasis on external relations, there is nothing to learn from utilitarianism concerning the intimate, generous, or friendly relationships that persons should develop toward other persons and with other beings in the world. Love, as described by Plato, and the truths that accompany a discussion of love are beyond the utilitarian's horizon. Despite these simple facts, as we have learned, few scientists see the limitations of the means-end fields of perception which underly much of their work. Consequently, many of the scientists who embrace the means-end approach, and establish fields of perception founded upon this approach, will not perceive important truths about life, including about their own life. Nor will they perceive major truths about human existence in the world, and its relationship to other beings.

We want to repeatedly emphasize a crucial perspective which reveals that the means-end approach is vacuous; the perspective relates to terms to which we have alluded. Those things that Plato called worthy in themselves, such as justice, wisdom, love, and beauty, have no place in a field of perception that is established upon utilitarian intentionality. Worse, by emphasizing utility as the basis for all human life and of all interaction with other beings, a utilitarian field of perception justifies and encourages irresponsibility toward the Being of these things that are worthy in themselves. Hence a fervent utilitarian does resemble Kierkegaard's seducer in one trait. Like Johannes, the fervent utilitarian probably would often relate cynically towards other human beings and towards those things that Plato called worthy. Need we add that utilitarianism and means-end approaches disregard and disparage spiritual existence?

Let us repeat the above point since its significance is profound. Once again let us compare the capitalist means-end field of perception, and the deeds that it seemingly justifies, to the field of perception that Shiva promotes in her books. If we look carefully, we discover an insight that philosophers have stated since the birth of utilitarianism which is constructed on means-end fields of perception. The assumptions that are at the basis of utilitarian intentionality, these thinkers have stated, block all truths about justice, wisdom, love, and beauty. Many utilitarians and capitalists even attempt to relegate these truths to oblivion. Shiva can write about these worthy matters; utilitarians cannot write about these matters without betraying their assumptions.

The insight has a corollary. In order to block these worthy truths and the fields of perception in which they come into being, the means-end approach must conceal many additional simple truths. One such simple truth is that we live with many other natural species in the world, and share the world with them; moreover, our existence and the existence of the world as we know it is bound to the well-being of these species. Geology and natural history teach us that many of these species evolved upon the face of the earth during many thousands and millions of years. Now, due to human indifference, irresponsible intervention, and ruthless greed, many species are threatened with extinction.

Thus, it is our responsibility to see this simple truth and to act in accordance with it. One immediate consequence of seeing this truth is that we should care for the well-being of these species. Moreover, it is our responsibility to establish fields of perception in which this simple truth is evident.

It would not be an exaggeration to state that the ruin of the environment continues at a ruthless pace because many capitalists, scientists, and bureaucrats refuse to comprehend and to act in accordance with the previously mentioned simple truth. They decide to be myopic to the fact that our existence, and the existence of the world as we know it, is bound up with the well-being of other natural species. They refuse to alter their myopic fields of perception which justify their daily deeds, many of which are wicked deeds. We can conclude that, by blinding their eyes to the simple truth that Vandana Shiva and many conscientious environmentalists, such as Rachel Carson, have articulately presented, these capitalists, scientists, and bureaucrats are acting irresponsibly. Furthermore, they are doing evil deeds and promoting evil.

But these irresponsible persons, and the fields of perception that they promote, have widespread support within the capitalist establishment, and also among ruling politicians. Hence, they are rarely criticized by the mainstream media. The results are quite evident. In democracies, it is difficult to find mainstream realms of public debate in which to attack and reject the gross irresponsibility that is spread by the princes of corporate capitalism and by their many supporters, among them academic lackeys. Part of the lore of the environmentalist movement describes the difficulties that Rachel Carson faced when she attempted to find a publisher for *Silent Spring*. From this lore we conclude that many publishers did not want to challenge the accepted fields of perception which supported corporate capitalism and, with it, the deeds of the United States government. From this perspective, we return to an example already presented in Chapter 6. The example reveals the deceit and the irresponsibility that prevails when scientists, working together with politicians, corporate capitalist companies, and government bureaucrats, block truths and spread mendacity.

Look again at the words of L. O. Howard, chief of the United States Bureau of Entomology in the 1920s, which were presented in a published speech. As pointed out in Chapter 6, Howard's statement is taken from a book by Edmund Russell which describes the war that human beings have declared upon insects and nature.[6] In his statement, Howard said clearly that we human beings must exterminate the entire class of insects that exist in the world. We already pointed out that the statement is stupid, since human existence would vanish if there were no longer insects—as has been shown countless times. Indeed, one of the reasons that we repeat Howard's statement is because it is stupid, and because, despite this stupidity, it was hardly ever challenged by writers in the mainstream press. We can now state that its stupidity was concealed by mendacious fields of perception promoted by capitalists in chemical companies and by other persons, among them politicians and government employees. In short, it seems that the inanity of Howard's statements was concealed by persons who would profit from his stupid statements, and by an irresponsible media and scientific milieu.

Here we want to emphasize that Howard's statement is irresponsible. Furthermore, the fact that the statement was not attacked, straightforwardly and immediately, by scientists merely adds to the atmosphere of irresponsibility toward the beings of nature that, according to Russell's study, prevailed in the United States for decades. Recall that L. O. Howard was an entomologist, chief of the United States Bureau of Entomology. Hence he knew the truths about insects, and his statement is both irresponsible and mendacious. It purposely conceals truths about the major benefits that insects bring to humanity and to life upon earth. It ignores the well-known truth about the links between human existence and the existence of insects and other classes of living beings that live upon the earth.

Furthermore, note that the field of perception that seems to underly Howard's statement is utilitarian. It is the field of perception of a conqueror, of a rampant exploiter of the resources of the earth. As mentioned, such is a field of perception established by the people who consider themselves to be lords of the earth. Within such a field of perception, blended with mendacity, it is not surprising that stupidity thrives.

We repeat this major point. Howard's proffered field of perception concerning the class of insects, which is stupid, includes falsehoods blended with hubris and arrogance. It is a field of perception which encourages human beings to live irresponsibly on the face of the earth. Unfortunately, as Edmund Russell has shown in detail, Howard was hardly exceptional in establishing a field of perception in which inane intentionality blended together falsehoods, hubris, and utilitarianism—a field of perception that encouraged rampant destruction of the environment. Even today, in the twenty-first century, we continue to encounter many such inane and destructive fields of perception spread by irresponsible scientists and bureaucrats, who work together with greedy capitalists and pernicious politicians.[7]

In summary, on the basis of Merleau-Ponty's insights it is evident that one of our basic responsibilities, as human beings who live upon the earth, is directly related to the fields of perception that we establish. Sensitive and truthful fields of perception, such as those established by Vandana Shiva, Rachel Carson, Jose Bove, and other strugglers for the environment can lead to responsibility for other beings with whom we share the earth. Persons who accept, adopt, and spread utilitarian fields of perception usually live irresponsibly upon our planet. Moreover, people who embrace and live with mendacious and inane fields of perception often live the life of a spiritual dwarf.

Even without reading Merleau-Ponty, some of those persons who struggle to save the environment grasp that their responsibility begins with the attempt to change accepted fields of perception. *Silent Spring* was Rachel Carson's attempt to change the field of perception of her readers about the poisons against insects that were irresponsibly sprayed in many areas of the United States. Jose Bove's ongoing struggles in France against industrialized agriculture have also changed many persons' fields of perception. Bove has shown that we can adopt other fields of perception in relation to agriculture—fields of perception that do not relate exploitatively to the farming of the soil, and that reject the pernicious actions of capitalist farming conglomerates that ruin nature.

Like Rachel Carson and Jose Bove, many environmentalists recognize that one way to counter the pernicious and mendacious fields of perception that prevail in the world is by establishing alternate fields of perception—fields that disclose major truths about our world. In addition, these strugglers for the environment recognize that they must encourage other people to perceive the truths that emerge within these fields. Carson, Shiva, Bove, and many others dedicated many years of their lives to such a worthy endeavor.

Yes, the fields of perception proclaimed and taught by Vandana Shiva, Rachel Carson, Jose Bove, and by other noble strugglers for the environment bring forth truths that have been concealed in many mainstream fields of perception. To borrow a phrase from Heidegger, by establishing these fields of perception, Shiva, Carson, Bove, and their noble colleagues have been able to open a hidden realm and to bring forth major truths from concealment. Frequently, these unconcealed truths—with all the horror and the splendor that they bring forth—enlighten many human beings.

NOTES

1. Soren Kierkegaard, "Diary of the Seducer," *Either/Or Volume I*, trans. David F. Swenson & Lillian Marvin Swenson (Princeton: Princeton University Press, 1959), 297–440.

2. See Maurice Merleau-Ponty, *Phenomenology of Perception*, trans. Colin Smith (London: Routledge & Kegan Paul, 1962), 456.

3. Vandana Shiva, *Stolen Harvest: The Hijacking of the Global Food Supply* (London: Zed Books, 2000), 1.

4. For the evils of the scientists and the capitalists of Monsanto, see Shiva, *Stolen Harvest*. Also, Robert Ali Brac De La Perriere & Frack Seuret, *Brave New Seeds: The Threat of GM Crops to Farmers* (London: Zed Books, 2000). Many other books on saving the environment describe the destruction brought about by the scientists and the activities of Monsanto.

5. For another example of such a field of perception, see: Vandana Shiva, *The Violence of the Green Revolution* (London: Zed Books & Third World Network, 1991).

6. Edmund Russell, *War and Nature: Fighting Humans and Insects with Chemicals from World War I to SILENT SPRING* (Cambridge: Cambridge University Press, 2001), 76–7.

7. For an example of a critique of such destructive fields of perception adopted by many scientists, see: Mae-Wan Ho, *Genetic Engineering, Dream or Nightmare: The Brave New World of Bad Science and Big Business* (Penang, Malaysia: Third World Network, 1998). For additional critiques in this area, consult journals such as *Seedling*, published by Genetic Resources Action International; *Genewatch*, published by Council for Responsible Genetics; *Global Pesticide Campaigner*, published by Pesticide Action Network.

Part Three

Sharing the Earth

8

Sharing the Earth as a Whole Person

In previous chapters, we frequently suggested that Merleau-Ponty's phenomenological description of perception has relevance for one of the most urgent challenges that we human beings face today. The challenge to which we referred is that we human beings must share the earth with other species and other beings. We must stop relating to other beings that are found in the world as existing there for us to dominate, conquer, and exploit. If we do not face this challenge straightforwardly, most environmentalists hold, we will bring devastation to much life on earth, including to the life of human beings.

This urgent challenge received ontological support in the discussions that followed upon our descriptions of Merleau-Ponty's findings concerning perception. These discussions were based upon the major insight that we are bodily involved and engaged in the world in every act of perception. Furthermore, every act of perception opens a field of perception in which the being that is perceived exists on a background. Hence, the ruining of the environment destroys the world as we perceive it and are involved in it through the fields of perception that we establish. In our discussions, we repeatedly stated that we human beings should relate to other species, and to the environment, with modesty and responsibility. We should relate as beings who are able to think and care, and who share the world with these many other species and beings. We also indicated, albeit briefly, how and why sharing the world with other species and beings was significant for a worthy human existence.

Looking back at the previous chapters, we note that underlying our discussions is an assumption that we have not yet straightforwardly stated. The assumption has to do with the way of life of the perceiving person, as described by Merleau-Ponty, and with the fields of perception that a person may intentionally establish. The time has come to state and elucidate this previously unstated assumption.

Put succinctly, our discussions have tacitly assumed the following: A life in which a person decides to share the earth with other human beings and with many other living species, and acts in accordance with this decision, frequently leads to a more worthy and whole human existence.

The assumption requires that we define what we mean by "a worthy and whole human existence." Note that there are many statements that relate to this topic in previous chapters. We shall present such a definition shortly; in that definition we learn from philosophers who related to the question of a worthy and whole human existence. First, however, we whisper a word of caution. The assumption that we have formulated should not be grasped as part of a moral theory or as suggesting a categorical imperative. The assumption emerges primarily from Merleau-Ponty's ontology of perception; as such, it has no primordial link to ethics. We repeat: the above-formulated assumption is a result of our thinking about the implications of Merleau-Ponty's ontology of perception. It has no direct link to ideas and concepts that are prominent in theories of morality, such as Kant's categorical imperative or kingdom of ends.

Consider a sentence from *Phenomenology of Perception* which partially contributes to clarifying our assumption. The sentence also helps us to point to the link between a definition of a worthy human existence and Merleau-Ponty's ideas on perception.

The relations between things or aspects of things having always our body as their vehicle, the whole of nature is the setting of our own life, or our interlocutor in a sort of dialogue.[1]

Merleau-Ponty's ideas in this citation are based upon the ontology of perception that he presented. This ontology led Merleau-Ponty to hold that our perception establishes "the whole of nature" as "the setting of our own life." Furthermore, our perception determines nature as "our interlocutor in a sort of dialogue." From the citation we also learn that our bodily perception on the pre-reflective level, and our bodily relations to things that we perceive, establish us as beings that can relate "in a sort of dialogue" with "the whole of nature." Hence, the citation reveals major truths about our primal existence. We believe that these truths point to a direction for a worthy and whole life. We shall explain the foundation of our belief while pointing to elements of a worthy and whole human life.

In previous chapters, we indicated that we accept the definitions of many of the philosophers of the past concerning a worthy life. Such a life, Plato and other thinkers since ancient Greece have held, requires courage and is dedicated to the disclosing of truth and to the pursuit of justice and wisdom. A well-known example of a person in ancient Greece who lived a courageous life dedicated to the disclosing of truth and to the pursuit of justice and wisdom is Socrates. His dedication to such a worthy life and his mission as an educator is articulately described by Plato in his seventh letter and in his Socratic dialogues.

Let the reader now read Plato's Socratic dialogues, say *Phaedrus* and *Symposium*, with the above citation from Merleau-Ponty in mind. The careful reader may discover that in these dialogues Plato, at times, describes Socrates relating to nature as the setting of his life. The reader may even find Socrates relating to nature as an interlocutor in a sort of dialogue. We repeat this point. The life of Socrates unites the courageous search for truth, the pursuit of wisdom and of justice with, at times, his relating to nature as the setting of his life and as an interlocutor in a sort of dialogue.

For instance, in *Phaedrus*, Socrates delights at the possibility of sitting on the banks of the Ilissus stream and bathing his feet in the cool waters of the stream, while listening to the shrill chorus of the cicadas coming from the trees that provide shade. He sees this small section of the natural world as an appropriate setting for him to engage in dialogue with Phaedrus on the nature of love and of rhetoric. He presents a myth that tells of the origin of the cicadas and illuminates some of his thoughts on rhetoric. Socrates's manners of relating to nature, we believe, are part and parcel of his choice to live as a whole person. We can now attempt to weave the threads of the discussion into a coherent pattern.

Before proceeding, however, we acknowledge that, when discussing Socrates's and Plato's lives and thoughts, we purposely skip the question of slavery as it existed in the Greek polis in the centuries in which the Greek genius flourished. Our views on this topic are resolute. Slavery is evil. Enslaving another human being is evil. We hold that all manners of human enslavement should vanish from the face of the earth. Yet, we also believe that the existence of slavery in Athens, where Socrates was nurtured, lived, and was active, did not have very much to do with Socrates's choices to seek to bring forth truths from concealment and to courageously pursue wisdom and justice. Any reading of Plato's Socratic dialogues will recognize that the existence of slavery in Athens probably had a minor influence on Socrates's way of life.

From Plato's dialogues we learn that Socrates's life accorded with our previously stated assumption. Socrates was a generous sharing person who endeavored to relate as a whole being to whomever and whatever he encountered. He relentlessly sought truths and wisdom and pursued justice. He probably sensed that such a search and pursuit must lead to the opening of fields of perception and to the sharing of his being with his contemporaries. In his relations with other beings and persons, we do not encounter Socrates relating as a conqueror or an exploiter or an oppressor, but rather as a generous person who shares his quest for new knowledge and wisdom, and even for new fields of perception with persons whom he encounters.

In some of the Socratic dialogues, Plato briefly shows that Socrates's fundamental attitude of relating as a whole being, and his sharing of his entire being with others, includes sharing the world with the beings of nature. We have already mentioned the myth of the cicadas that Socrates relates in *Phaedrus* before turning to a discussion of rhetoric. The myth relates the legend that human beings and cicadas have common ancestors. Hence, the myth can direct persons to learn to relate modestly toward beings of nature, and with a wish to share the world with cicadas and with other such beings.

Socrates's life also suggests that a person cannot seek truths, cannot be a lover of wisdom, and cannot pursue justice if he or she does not endeavor to act generously—without conditions and with one's whole being. A conditioned generosity, in which a person carefully reckons what generous acts are worth performing, is no generosity at all. We believe that almost all of contemporary capitalist philanthropy, including philanthropy through foundations, is a conditioned generosity. The same is true of much of the aid that wealthy countries bestow on impoverished countries—very often such is an act of conditioned generosity.

Any true act of generosity, Plato shows, has no strings attached. Relations of generosity may often require sharing one's entire being and thoughts with other persons. Such a sharing of one's being and thoughts frequently means making room in one's everyday existence for other persons, and at times, for other beings. Socrates constantly made room in his life for fellow Athenians who were willing to accompany him during his relentless search for truth and wisdom, and to share his fields of perception. Plato's Socratic dialogues frequently indicate that such a making of room in one's life for other persons is the act of a person's whole being, and that living as a whole being may lead to acting generously toward other beings that exist.

Deciding to share the earth with other beings, and living in accordance with this decision, is more than an act of generosity. It is a living fully of some of the truths concerning human existence that emerge in the ontology of perception that Merleau-Ponty described. Let us assume the truth of Merleau-Ponty's above statement that, in my bodily perceptions, nature is my "interlocutor in a sort of dialogue." This truth leads to some poignant questions. What is the "sort of dialogue" to which Merleau-Ponty refers?

What do persons express in this dialogue, and what does nature express? And most important for the theme of this book: Is there any reason for human beings to daily endeavor to bring ruin and devastation to this primal interlocutor, that is, to nature?

In this study, we cannot provide full answers to these questions. We shall point, however, to a direction in which, at least, partial answers can be found. In order to discern this direction, listen to Merleau-Ponty's explanation of our relation to other beings; the following appears a few sentences after the previous citation:

> The thing is inseparable from a person perceiving it, and can never be actually *in itself* because its articulations are those of our very existence, and because it stands at the other end of our gaze or at the terminus of a sensory exploration which invests it with humanity. To this extent, every perception is a communication or a communion. . . .[2]

We have already mentioned that, at times, Merleau-Ponty presents poetic locutions in order to clarify his ideas. Such occurs in this citation, for instance, in the last quoted sentence. Yet, these poetic locutions point to an aspect of reality that is frequently overlooked. Hence we should hark to the insights that they disclose.

We also recognize that in this citation there may be grave philosophical problems in some of Merleau-Ponty's assertions. We are especially concerned about the assertion concerning the status of the thing as an in-itself. However, we purposely skip the problematic assertions in the citation, since they do not diminish Merleau-Ponty's illuminating ideas, to which we turn.

One enlightening idea in the above citation is that in every perception, by the act of perception, I invest every thing that I perceive with what Merleau-Ponty denotes "humanity." This investment of a person's humanity in the thing perceived is central to the "sort of dialogue" that may emerge between beings of nature and the perceiving person. This investment of humanity in perception is a component of what Merleau-Ponty calls a person's communication and communion with the beings that he or she perceives.

In order to clarify this idea, consider two simple examples. On a hike on one of the less traveled trails of a national park, I reach a trailhead, perceive a trail that climbs, and judge it "too steep." By this immediate judgment, I am investing this trail that climbs into the mountains with my humanity, with the situation of my being. I am tired and perhaps a bit of a coward, hence I decide that the trail is "too steep." I turn away and follow another trail. The humanity that I invested in the trail that I called "too steep" partially expressed my own fatigue and fears. It would be difficult to hold that I conducted a "sort of dialogue" with the steep trail. I judged it, turned away, and chose another trail.

Later on the hike, I reach a lovely, small, glimmering glacier lake surrounded by mountains and white snow fields. The lake's deep blue and tranquil loveliness captures my imagination; joy wells up in me as I sit and let my gaze sink into the blueness of its clear waters. Slowly, I feel that the joy welling up in me while I contemplate the lake's beauty soothes my body. There are moments when I want to share this experience of blue tranquil loveliness with other human beings.

The two examples disclose that there is much truth in Merleau-Ponty's statement that the act of perception always invests the beings that are perceived with the perceiving person's humanity. The examples also reveal that this invested humanity may be reflected from the being that is perceived. When I immediately choose to turn away from a path that is "too steep," I have hardly engaged in dialogue with it. In the case of the deep blue glacier lake, a sort of dialogue may emerge, thanks to the humanity that I invested in the perception of the details of its beauty. In short, Merleau-Ponty's poetical locutions relate to existential situations that emerge in everyday life.

Additional examples which confirm these poetical locutions can also be found in the writings of great authors. A classic example of an essay by a person who invests his humanity in beings of nature, while living with them and perceiving them carefully, is Henry David Thoreau's autobiographical essay *Walden*.[3] In this essay, Thoreau frequently described the "sort of dialogue" that emerged in his relations with the beings of nature that he encountered when he lived alone in the woods in a hut by Walden Pond. He also articulately describes the rich fields of perception that he established during this period, fields of perception which disclosed much about the natural beings that he encountered.

Among the more recent literary examples of a sort of modest and responsible dialogue with beings of nature is Ernest Hemingway's *Green Hills of Africa*.[4] (We already discussed the dialogue between the fisherman Santiago and a fish that is found in Hemingway's novel *The Old Man and the Sea*.) This autobiographical description of a hunting trip in Africa portrays Hemingway's perceptual investment in many beings of nature that he encounters while hunting; it also describes the perceptual fields that he established and the sort of dialogue that he conducted with many beings. In both Thoreau's and Hemingway's autobiographical essays, the reader will not find it difficult to encounter descriptions of the authors' perceptions which are moments of communication and communion with beings of nature.

In contrast to my encounter with the small glimmering glacier lake, or to the experiences with nature that are described by Thoreau and by Hemingway, consider again those people who dedicate their lives to being conquerors, exploiters, or oppressors. The major goal that these persons wish to realize in their life is to rule and to exploit other human beings and other natural beings that exist in the world. Such a conqueror, exploiter, or oppressor will tend, while perceiving beings in the world, to block what Merleau-Ponty referred to as a "sort of dialogue" that may come into being with things and with nature.

For instance, usually the conqueror, exploiter, or oppressor of other human beings will not engage in dialogue with the oppressed, exploited, or conquered persons that are given to his or her rule. Indeed, we do not need to rely on Merleau-Ponty's thinking to recognize that in such a person's relations and perceptions, there is an ongoing attempt to block dialogue—any dialogue. Any encounter with persons who live as conquerors, exploiters, and oppressors usually discloses that they block dialogue with all other beings, among them beings of nature. Such a blocking of dialogue may greatly hinder a person's ability to live and to relate as a whole being. In addition, a person who blocks dialogue with other persons and with beings of nature will usually develop limited fields of perception.

Here someone may ask: What would be the aspects of humanity that the conqueror invests in the people he or she has vanquished? Our answer is linked to ontology. Being human frequently requires that a person decide to submit his or her being and freedom, at least partially, to overwhelming external powers. Among these external powers are what may be termed "the powers of nature," say, a severe monsoon or a hurricane. The conqueror's choice of fields of perceptions and his or her actions may convey the attitude that the vanquished must live with a day-to-day choosing of submission of their being and freedom to the conqueror's rule as if it were an overwhelming external natural power. Such a confining of one's own fields of perception to domination without sharing, as is done by the conqueror, exploiter, or oppressor, attempts to block any possibility that a perception may emerge that is "a communication or a communion" with other beings.

We have already provided the information needed to answer an important question that is linked to the being of the exploiter and conqueror. The question is: What happens when a person decides to deliberately block central elements that may emerge in his or her perceptions? Based on what we have discussed, here and in previous chapters, the immediate answer is that such a person also chooses to establish and to live with confined and shallow fields of perception. And living with confined and shallow fields of perception condemns that person to ignorance, narrowness, and myopia. It also blocks the possibility of living as a whole person.

The blocking of elements of perception in order to establish straitened and superficial fields of perception is hardly a rare phenomenon. In Chapter 6, we already indicated that millions of impoverished people, who daily struggle for bodily survival, develop only limited fields of perception. We shall soon mention two works of literature that describe such a blocking of elements of perception by persons who belong to the upper classes. Unfortunately, as these and other works of literature suggest, the world is swamped with people who deliberately choose to establish and to live in straitened and superficial fields of perception. The perceiving of beings within these confined shallow fields of perception hinders communion with many natural things. Frequently, living within these confined and shallow fields of perception excludes and hinders any "sort of dialogue" with beings of nature, and even with other human beings. Such people have emasculated their opportunity to exist as whole persons.

How does life in society contribute to the establishing of these confined fields of perception? As described in Chapter 6, harshly exploited and enslaved persons, whose major everyday concern is mere survival, will tend to develop confined and shallow fields of perception. However, as already stated, many members of the elite of a society choose to limit their fields of perception. Hence, we do not believe that the choice of straitened and shallow fields of perception is necessarily linked to poverty or to those human beings who are called "the members of the lower classes." We do believe that many members of the upper classes flee from the possibility of opening new fields of perception. Put differently, upper-class society rarely encourages the opening of new fields of perception, and the living as a whole being.

Look again at Marcel Proust's *Remembrance of Things Past*. Proust has shown in enlightening detail how the so-called noblest elements of Parisian society—those Parisians who delighted in the wit of the Guermantes—daily inculcate their members, and those who aspire to join their circle, to embrace and to live with shallow confined fields of perception.[5] Thus, from Proust, and from other great authors such as Tolstoy, we learn that confined and shallow fields of perception are found, and are often encouraged, in the perceptions of members of all social classes and among adherents of all political views. We also learn that breaking out of these confined and shallow fields of perception, fields that social life tends to impose upon the members of a specific society, is frequently a difficult and courageous personal decision. Yet, without such a courageous decision, a person will not live as a whole being who frequently opens and relates to new fields of perception.

Works of literature also reveal that, often, those persons who deliberately narrow their field of perception find themselves choosing to exist in a vicious circle of mediocrity and banality. Count Oblonsky, as he is described in Tolstoy's novel, *Anna Karenina*, is such a person.[6] Oblonsky's continual choice of straitened and shallow fields of perception determines that he live a mediocre existence. Tolstoy shows that these narrow fields of perception do not allow many aspects of Oblonsky's humanity to be expressed. Thus, Oblonsky and his many mediocre friends cannot invest many things, for instance works of art or beings in nature, with their own humanity. These members of the decadent Russian nobility do not live as whole beings. In turn, the mediocre way of life embraced by Oblonsky, and by many other of his fellow Russians, leads them to repeatedly choose straitened and shallow fields of perception. They thus choose to be stuck in their mediocrity and banality.

The results of such a vicious circle of mediocrity and banality are sad and sordid. Those persons who live daily in a vicious circle of mediocrity do not experience perception as a communication or a communion. Nor do they exist as whole persons who can courageously decide to open new fields of perception and to share them with others. Instead, Oblonsky, and those who resemble him, merely want to exploit other beings that exist in the world for their own shallow pleasure and enjoyment.

We repeat. The possibility of a mediocre and banal person, such as Tolstoy's Count Oblonsky, broadening his or her fields of perception is continually obstructed by his or her daily banal choices. When a person chooses to be trapped in such a vicious circle of mediocrity and banality, he or she will not find ways to relate as a whole being to other human beings. Nor will such a person relate as a whole being to the beings of nature. These persons, many authors show, will never experience a moment of relating wholly, such as the moment described by Erich Maria Remarque, with which we close this chapter.

We want to again emphasize that Merleau-Ponty's phenomenology of perception indicates that there is a possibility of transcending the vicious circle of mediocrity and banality that is central to the life of many persons. There is a possibility to live differently. An anti-mediocre life will include a constant attempt to enlarge one's fields of perception. And this directing of oneself to broadening one's fields of perception may lead to living fully, as a whole person. We hold that such a life may also include beautiful moments of sharing with other beings, including beings of nature. As mentioned, the reader of *Walden* and of *Green Hills of Africa* will encounter two authors, Thoreau and Hemingway, who endeavored to enlarge their fields of perception while relating to natural beings. These authors also experienced moments of sharing, of communication and communion, with beings of nature. Frequently, they both strove to relate as whole persons to the natural beings, and to the human beings, whom they encountered.

Because we believe that living as a whole person is spiritually enhancing, we want to add some depth (and beauty) to the previously mentioned ideas that stem from Merleau-Ponty's findings. We shall do so by closing this chapter with a moving literary example which shows the beauty of living fully, as a whole person, who is always willing to enlarge one's fields of perception. The example shows that being fully alive, with broadening fields of perception, with perceptions that are a communication or communion, often leads to joy in sharing the earth with other beings. The reader should not find it difficult to link this moment of sharing of the earth with additional ideas that we have presented from Merleau-Ponty's ontology of perception.

The literary example which we shall bring describes the happenings of an evening and a night in the life of Dr. Ravic, as they are presented in Erich Maria Remarque's novel *Arch of Triumph*. During this summer evening and night, Ravic lives fully, as a whole person, while perceiving broad fields of perception and sharing the world with natural beings. Remarque shows that such a moment of sharing can include a rejuvenated love of life. As such it is a moment of spirituality, and has no need of comment. We have decided to close the chapter with Remarque's description of that moment.

The hero of the novel, Dr. Ravic, is a courageous person and an excellent surgeon. Ravic has fled Nazi Germany after he was tortured for hiding two friends who had opposed Nazism in his house. The friends were caught and killed. Since 1934, Ravic has been living and working illegally in Paris. The novel opens in late 1938; already many people in Europe know that war between Germany and other European nations is inevitable.

One wintry night, Ravic met Joan Madou, a nightclub singer and minor actress, reeling in the street. He helped her overcome a deep personal crisis. Later they became lovers. But when Joan Madou learned of Ravic's illegal status and its consequences, she slowly drifted away, living with men who were not in France illegally, and who could financially support her. Despite his deep love for Joan, Ravic cuts off all relations with her. In the early summer of 1939 Ravic experiences the illuminating moment that we wish to cite.

Late one summer night, after a long, exhausting evening walk through the streets of Paris, Ravic finds himself sitting in a garden near Joan Madou's high apartment building. The rent for Joan's apartment has been paid for by another man. He looks at the light in her high apartment and feels a deep pain. But he knows that, even if Joan returned to him, she could not return wholeheartedly. The hour for true love between them to flourish has vanished. Suddenly Ravic hears heavy thunder.

Raindrops splashed on the bushes. Ravic got up. He saw the street mottled with black silver. The rain began to sing. The heavy drops beat warmly against his face. And suddenly he no longer knew whether he was ludicrous, or miserable, whether he was suffering or not—he only knew that he was alive. He was alive! He was there, it held him again, it shook him, he was not a spectator any longer, not an onlooker from the outside; the great splendor of uncontrollable feeling shot through his veins again like fire through a furnace; it scarcely mattered whether he was happy or unhappy, he was alive and he was fully aware that he was alive and that was enough.[7]

Ravic turns toward the light in Joan's apartment, peers at it, and speaks. In a long soliloquy, he thanks her for helping him return to being fully alive in a world plagued by catastrophe, much evil, and widespread misery. By his experiencing the anguish and pain of his failed love with Joan, Ravic recognizes that he can live fully; he is alive. After the soliloquy, soaked to the skin and calm, Ravic turns to go. He perceives Paris in the rain.

The rain had turned into a glistening silver curtain. The bushes became fragrant. The smell of the soil was strong and grateful . . . The night was there shaking rain down from the stars; mysterious and fructifying it poured down on the stone city with its alleys and gardens, millions of blossoms held out to it their multicolored sex and conceived it; it flung itself into the millions of open arms of the trees and penetrated the soil for its dark nuptials with millions of waiting roots, the rain, the night, nature, growth, they were there, unconcerned about destruction, death, criminals, false saints, victory or defeat, they were there as in every year, and on this night he belonged with them; the shell had broken open, life stretched out, life, life, life, welcomed and blessed.[8]

NOTES

1. See Maurice Merleau-Ponty, *Phenomenology of Perception,* trans. Colin Smith (London: Routledge & Kegan Paul, 1962), 320.
2. Ibid.
3. Henry David Thoreau, *Walden and Civil Disobedience* (New York: Signet, 1960).
4. Ernest Hemingway, *Green Hills of Africa* (London: Jonathan Cape, 1936).
5. Marcel Proust, *Remembrance of Things Past,* trans. C. K. Scott Moncrieff (New York: Random House, 1934). See especially the section "The Guermantes Way."
6. Leo Tolstoy, *Anna Karenina,* trans. David Magarshak (New York: New American Library, 1961).
7. Erich Maria Remarque, *Arch of Triumph,* trans. Walter Sorell & Denver Lindley (New York: Ballantine Books, 1998), 306.
8. Ibid., 307.

9

Honest Engagement in Sharing the Earth

In this concluding chapter we shall stress some ideas that have been presented, and add a dimension to their significance. First among the ideas that underlie our study is the fact that, among their other contributions to philosophy, existentialist thinkers in the twentieth century revived the study of ontology. As indicated, both Martin Heidegger and Jean-Paul Sartre showed that ontology, that is, thinking about the Being of beings and about Being, is crucial for relating to and comprehending human existence. They also pointed out that ontology is crucial for the challenge of authentically assuming responsibility for the world in which we exist. Maurice Merleau-Ponty learned much from both philosophers.

Already in the Introduction we mentioned that in *Being and Time* Heidegger describes the human entity, Dasein, as Being-in-the-world. We have mentioned his statement that Dasein is also characterized by its ability to care for other beings. Hence, Dasein as a caring being, is at least partially responsible for the other beings that it encounters in the world. In his profound study, Heidegger also deals with specific descriptions of Dasein's various modes of existence, many of which are not worthy. In this context, he presents the ontology of Dasein's inauthenticity. In one of the sections of *Being and Time,* he explains in detail that inauthenticity emerges when Dasein allows the "they" to determine its everyday life and decisions. In other chapters, we mentioned some of the worthy challenges that Heidegger presented to human beings; we stated that he believes that the greatest challenge for Dasein is to be the shepherd of Being.

In *Being and Nothingness,* and in other studies and literary works, Sartre showed that human beings are condemned to be free. One result of this freedom is that persons who authentically live their freedom can relate responsibly to other beings in the world. In his study, Sartre also described in detail the ontology of bad faith and the various patterns by which bad faith emerges in daily human interactions. One instance of bad faith is the flight from responsibility for one's life and decisions. We dwelt briefly on these thinkers throughout this book because their work served as a background and an influence on Merleau-Ponty's writing.

We may criticize some of the ideas of the existentialist thinkers, or their way of life, but one truth emerges. The revival of thinking about Being, about the Being of beings, and about the ontology of human existence requires a philosophical honesty that, unfortunately, many other twentieth-century thinkers do not display. Expanding upon this point would probably require writing a book; hence we shall only discuss this dishonesty in passing. Already it is clear, even from our brief review of ideas in Merleau-Ponty's *Phenomenology of Perception,* that a philosopher who deliberately ignores one's own daily bodily existence and involvement in the world is not writing in good faith. Furthermore, a philosopher who ignores the fact that he or she is a free person who daily determines one's way of life, is not writing in good faith.

We should add that honesty concerning human existence is found in Heidegger's philosophical thinking—even if he was dishonest in his responses to political parties and to political developments. He seems to have been especially dishonest, as many scholars have shown and discussed, in his relationship to Nazism.[1] Despite these personal failings, Heidegger's honesty in thinking about human existence, about Being and the being of beings is sorely lacking in the writings of many other so-called twentieth-century thinkers.

As previous chapters have revealed, Merleau-Ponty's *Phenomenology of Perception* is written in the existentialist tradition. Even from the partial discussion of themes in the *Phenomenology of Perception* that was presented in this book, it is clear that Merleau-Ponty undertakes a philosophical challenge that requires thinking in the realm of ontology. We have shown that, already in his rejection of the tenets and so-called findings of empiricism and rationalism, Merleau-Ponty brings forth from concealment significant truths concerning the ontology of human perception. For instance, he shows that the constancy hypothesis is false and that we establish fields of perception within which we perceive the beings that we encounter.

In previous chapters we have indicated, albeit briefly, that the truths concerning human perception that Merleau-Ponty revealed have much to teach us. We have concentrated upon the fact that many of the truths concerning perception that Merleau-Ponty brought forth from concealment have worthy implications for our relationship to other species of nature and to other beings in the world. In addition, we have explained briefly that Merleau-Ponty's findings and thoughts concerning perception can indicate how to live a whole and worthy life while sharing the world with other beings, and, at times, caring for these beings. We admit that, while being faithful to many of Merleau-Ponty's ideas, we have gone beyond his specific findings and ideas. Put differently, we have shown new implications for human existence in the world on the basis of some of the truths concerning perception that Merleau-Ponty disclosed.

Such a going beyond the formulated thoughts of an existentialist thinker is not unprecedented. For instance, any thoughtful reader will discern that Heidegger's detailed descriptions of inauthenticity can lead a person to attempt to live authentically—say, not to give oneself to idle talk, curiosity, and ambiguity. Such an attempt to live authentically, we believe, may proceed in directions that Heidegger did not outline. Similarly, Sartre's descriptions of bad faith can lead persons to explore the possibility of living in good faith—a topic which Sartre did not discuss at length in his philosophical writings, although hints about living in good faith appear in his literature and drama.

In this vein, previous chapters have indicated that Merleau-Ponty's discussion of the ontology of perception and his ideas concerning the fields of perception can point to ways by which we may strive to enhance human existence and live responsibly in the world. We have suggested that one crucial area of such enhancement of human existence and of responsibility is our relationship to beings of nature and to other species. Let us be precise. We have repeatedly stated that Merleau-Ponty's findings concerning perception and fields of perception can serve as the basis for an honest engagement in sharing the earth with other human beings, and with other natural species. Such an engagement in sharing the earth, we have pointed out, is a manner of pursuing justice, and may frequently lead a person to live wholely.

Yet, given the state of philosophical thinking that prevails today in much of academia, we believe that it would be wrong not to mention, in this closing chapter, the importance of honest philosophical engagements in the world. We shall do so by briefly pointing to the widespread dishonesty that you may encounter in much contemporary philosophical writings. Such dishonesty, we believe, contributes to the many evils that we daily confront in the world, among them the ongoing destruction of the natural environment. The reader may already have guessed that we have a bone to pick with many postmodern and analytic philosophers.

Here is our major complaint. Many of today's so-called postmodern thinkers relegate truths to the realm of the relative. Such an approach constitutes a spitting in the face of philosophy as it has been conceived and developed since its inception. After all, any reader of Plato's Socratic dialogues will discern that Socrates endeavored to think honestly and to live his everyday life inspired by a firm commitment to truth—he repeatedly rejected any attempt to suggest that truth can be relative. Such was the approach adopted by all great Western thinkers—since the pre-Socratic philosophers.

It may seem obvious to repeat the principle that intellectual honesty requires that philosophers be committed to truth as truth, and that truth is not relative. Especially since, as many pupils in elementary school already know, honesty vanishes when truth is relative. Yet, from our readings of quite a few postmodern writings, we are no longer sure that the commitment to truth is obvious to many—if not most—of these so-called thinkers.

We are quite confident that many capitalists and their supporters are pleased when truths are considered to be relative. It allows them to hold intolerable views, as they have done in a roundabout manner while discussing, say, the recent history of Central America and of Suharto's Indonesia. Some of these supporters of capitalism have intimated that harsh exploitation of peasants is necessary for growth, and killing opponents is necessary for freedom. Such statements can be made only if one believes that truth is relative. Note that, as Noam Chomsky has shown in many books, the United States government usually supported such blatant deceit—frequently by distortion of truths and, at times, under the banner of relative truths.[2]

Previous chapters have indicated that this book is based on the principle that deciding to share the earth with other beings requires intellectual honesty, and a respect for truth. Such honesty will prevail, we hold, only if philosophers and laypeople think honestly, and dedicate their daily lives to bringing forth truth from concealment.

While working on this book, we noticed an interesting historical fact. All the trends of postmodern thinking—postmodernism is a so-called philosophical approach that emerged in the last four decades of the twentieth century—became popular in the period when the dimensions of the ecological catastrophe that human beings brought about by their daily decisions and deeds became evident. In a word, postmodern writers became acceptable in the popular mind in the years after Rachel Carson published *Silent Spring* in 1962.[3] Yet, these postmodern writers almost never related to the ongoing natural catastrophe, instigated by human beings, that is articulately described and condemned in *Silent Spring* and in other environmentalist writings. Moreover, they rejected the wisdom of existentialist thinkers whose writings were quite popular during this period.

We believe, but shall not try to prove, that the historical fact is not a coincidence. We hold that, by rejecting existentialism and ignoring the destruction of the natural world, these postmodern thinkers or, to borrow a term from Heidegger, these pseudo-philosophers, were choosing to narrow their fields of perception and to choose a shallow existence. They then endeavored to justify their choices in so-called philosophical terms. Such a dishonest choice may have brought them some benefits in twentieth-century capitalist society, but it gravely impoverished their own thinking. It also blocked all quests for a spiritual existence—which must relate to truths.

Furthermore, if the postmodern and analytic thinkers did not sense this impoverishment of their thinking, they were not true thinkers. Indeed, these pseudo-thinkers, for instance Jean-Francois Lyotard, presented many writings, which are hardly worth discussing for two reasons: they are intellectually dishonest and they are narrow and confining.[4] Again, as every child knows, dishonesty rarely leads to the bringing forth of truths.

But, someone may say, dishonesty in the history of philosophical thinking is hardly new. Such dishonesty is at least as ancient as Plato's dialogue *Gorgias*. We grant this point. In *Gorgias*, Polus and Callicles are described as dishonest thinkers; Polus, for instance, plays with words and is not willing to think about essences. Sophistry and rhetoric were his way of life. But, the fact that Polus was dishonest does not leave contemporary dishonest pseudo-philosophers off the hook. And the fact that history is populated with dishonest pseudo-philosophers does not relieve us from the responsibility of revealing the dishonesty of contemporary thinkers—albeit briefly.

Much of the dishonesty in contemporary philosophy that we are attacking can be described with the help of two key terms that postmodern writers continually use: "language games" and "narratives." During the past four decades, these terms have been embraced by postmodern and analytic thinkers, and, at times, by late twentieth-century pragmatist philosophers. Many postmodern thinkers have stated that narrative and language games are central to all philosophical thinking. One result of this statement is immediate. All truths, these thinkers hold, are relative; they belong to specific language games and are part of a certain social narrative. Note that none of these thinkers attempted to address the question of the essence of truth, as did, for instance, Martin Heidegger.[5]

In the context of our learning from the wisdom of Merleau-Ponty, the main problem with the terms "language games" and "narrative" is that they suggest that there are no unquestionable truths. Indeed, many, if not most, thinkers who promote these terms announce: all truths can be challenged and many accepted truths can be discredited. But Merleau-Ponty's phenomenology of perception assumes that truths exist.

Since these postmodern and analytic writers refrain from discussing the essence of truth, we can provide an argument that discloses the vacuity of their tenets. Although this argument is not crucial, it shows the inherent poverty of their so-called thinking.

In brief, these postmodern thinkers seem to have forgotten that their assumptions are groundless. After all, their own assumptions concerning the wide prevalence of language games and the significance of social narrative can be considered, in themselves, to be merely part of a social narrative and a language game. Consequently, the truth of their assumptions can be challenged and held to be merely part of a language game or a social narrative. We are then faced with an infinite regression which makes their assumptions and conclusions look dubious. Or as Heidegger would have put it bluntly, the assumptions presented by postmodern thinkers have no ground.

In order to gauge a few of the ruinous outcomes of these postmodern terms for honest philosophical engagement in the world, consider the challenge we human beings have of sharing the earth with other species. Let us assume that it is true, as environmentalists have stated and, at least partially proved, that, due to human actions, each year 10,000 (ten thousand!) species disappear from the face of the earth. One thing is evident. This true statement has nothing to do with language games or with narratives. It is a fact. Similarly, if as has been recently published, due to overfishing, ninety percent of the large fish in the world have disappeared from the oceans, this fact has nothing to do with a narrative or a language game. No social narrative or novel language game can bring back the fish that vanished or the thousands of species that became extinct.

We repeat. No discussion of language games or of narratives can challenge the truth concerning the disappearance of large fish from the oceans of the world. Flight from these and many other disturbing facts concerning the world, with the help of terms like "narrative" or "language games," is a flight from the truth. It is also a fleeing from the responsibility of human beings for the fate of the earth. Such is also the case in regard to a stark refusal—that emerges in the writings of postmodernists—to learn from Merleau-Ponty's findings about perception and the field of perception.

Phenomenology of Perception was published before both terms—"language games" and "social narratives"—became popular in philosophical circles. We have repeatedly shown that the truths concerning human perception that are found in Merleau-Ponty's book are existential, that is, they are central to human existence. They are ontological truths. As such, these truths have nothing to do with social narratives or with language games. Furthermore, these truths are not relative. Indeed, much as you cannot challenge the truth that Brutus stabbed and killed Julius Caesar, you cannot challenge many of the truths concerning perception that Merleau-Ponty disclosed. We can now add that refusing to relate to these truths, or calling them a mere language game or part of a society's narrative, is intellectually dishonest.

Thinkers may disagree with some of Merleau-Ponty's assumptions or they may challenge some of the truths and conclusions that he presents.

That is a legitimate philosophical undertaking. It is not an attempt to disparage his findings by relegating them to being part of a language game or a social narrative. Indeed, any honest reader of *Phenomenology of Perception* will discern that bringing forth truths about human perception—that is, the perception of all human beings—is not the presenting of a language game. Neither is it a narrative. Much as for Gertrude Stein the rose is a rose is a rose, we can say that in Merleau-Ponty's study of human perception, truth is truth is truth.

All this leads to a major conclusion. There are no relative truths in Merleau-Ponty's findings concerning human perception. Nor are there relative truths in the fact that human beings are daily acting so as to bring about the annihilation of thousands of natural species that populate the earth. We have primarily based our research and our writing of this book on these two sets of truths. This research and writing has shown that Merleau-Ponty's findings can lead to an honest engagement in relating to human perception, and in sharing the earth with other beings. This honest engagement can help us to establish and sustain a naturally diverse and exciting world for all human beings, and for the generations that will be born in the future.

For centuries, biblical scholars have faced a problem. God created light on the first day, but he created the sun, the moon, and the lights in the firmament of the heaven on the fourth day. Where did the light of the first days come from?

Talmudic scholars have suggested that the light of the first seven days—during which God created heaven and earth and all that lived on earth—was special. It was a Godly light that had nothing to do with the sun or stars; some scholars say that it was the light of truth. According to this view, the sun, the moon, and the stars only commenced to provide light in the world created by God after the seventh day. But, some scholars add, in every generation there are human beings who crave for the Godly light that is brought forth by truth.

In *Phenomenology of Perception*, Merleau-Ponty endeavored to illuminate significant and previously ignored truths concerning human perception. He succeeded. These truths, we have shown, can and should lead human beings to relate as sharers of the world with other natural species. Comprehending some of the truths that Merleau-Ponty brought forth from concealment, and discussing some of their implications for our relationship to other natural species that exist in the world, is the challenge and the engagement that we undertook in writing this book. We have attempted to show how truths concerning perception and fields of perception, that were presented by Merleau-Ponty, lead to and enlighten truths concerning daily human existence in the world. We then showed the manners by which these truths are especially relevant in developing an honest, whole, and worthy life. We particularly stressed that these truths lead to a life in which a person is engaged in the world as a being who shares the world with others, including beings of nature. By living such an engagement, persons assume responsibility for their existence in the world, and for the existence of other beings in the world, including the natural species that populate the earth.

In all due modesty, we are now willing to risk a courageous statement. The light emanating from the truths about perception and fields of perception that Merleau-Ponty presented can illuminate additional truths. Together, these truths can help bring forth a thinking about our manner of being in the world. We believe that such thinking can lead to a world in which justice prevails and in which persons care to share the world with other beings and species.

Will these truths be heeded?

NOTES

1. Heidegger's flirt with Nazism has aroused much scholarly discussion. See, for instance: Victor Farias, *Heidegger and Nazism,* trans. Paul Burrell & Gabriel R. Ricci (Philadelphia: Temple University Press, 1989); Richard Wolin, ed., *The Heidegger Controversy: A Critical Reader* (Cambridge, MA: MIT Press, 1993); Fred Dallmayr, *The Other Heidegger* (Ithaca, NY: Cornell University Press, 1993).

2. See, for instance, the chapters on these countries in Noam Chomsky, *Deterring Democracy* (London: Verso, 1991).

3. Rachel Carson, *Silent Spring* (London: Penguin Books, 2000).

4. Jean-Francois Lyotard, *The Postmodern Condition: A Report on Knowledge.* Trans. Geoff Bennington & Nrian Massumi (Minneapolis: University of Minnesota Press, 1989).

5. Martin Heidegger, "On the Essence of Truth," trans. John Sallis. The essay is found in Martin Heidegger, *Pathmarks* (Cambridge: Cambridge University Press, 1998), 136–54.

Selected Bibliography

Abram, David. "Merleau-Ponty and the Voice of the Earth." David Macauley, ed. *Minding Nature: The Philosophers of Ecology*. New York. The Guilford Press, 1996. pp. 82–101.
———. *The Spell of Sensuous*. New York: Vintage Books, 1997.
———. "The Perceptual Implication of Gaia." *The Ecologist*. Vol. 9, No. 2, 1987.
Ali Brac De La Perriere, Robert & Frack Seuret. *Brave New Seeds: The Threat of GM Crops to Farmers*. London: Zed Books, 2000.
Berdyaev, Nikolai. *Slavery and Freedom*. Trans. R. M. French. New York: Charles Scribner's Sons, 1944.
Bove, Jose & Francois Dufour, Interviewed by Gilles Luneau. *The World Is Not for Sale: Farmers Against Junk Food*. Trans. Anna de Casparis. London: Verso, 2001.
Burgos-Debray, Elisabeth, ed. *I Rigoberta Munchu: An Indian Woman in Guatemala*. Trans. Ann Wright. London: Verso, 1984.
Camus, Albert. *The Rebel*. Trans. Anthony Bower. New York: Vintage, 1991.
Carson, Rachel. *Silent Spring*. London: Penguin Classics, 2000.
Chomsky, Noam. *Deterring Democracy*. London: Verso, 1991.
Colborn, T. et al. *Our Stolen Future*. London: Abcus, 1997.
Cummings, David, ed. *Encyclopedia Dictionary of Classical Music*. New York: Random House, 1977.
Dallmayr, Fred. *The Other Heidegger*. Ithaca, New York: Cornell University Press, 1993.
Dostoevsky, Fyodor. *Crime and Punishment*. Trans. Richard Pevear & Larissa Volokhonsky. New York: Vintage Books, 1993.

Farias, Victor. *Heidegger and Nazism.* Trans. Paul Burrel & Gabriel R. Ricci. Philadelphia: Temple University Press, 1989.

Heidegger, Martin. *Being and Time.* Trans. John Macquarrie & Edward Robinson. Oxford: Basil Blackwell, 1962.

———. "Letter on Humanism." David Farell Krell, ed. *Martin Heidegger: Basic Writings.* New York: Harper & Row, 1977.

———. "On the Essence of Truth." Trans. John Sallis. The essay is found in Martin Heidegger, *Pathmarks.* Cambridge: Cambridge University Press, 1998. pp. 136–54.

———. "The Question Concerning Technology." In *The Question Concerning Technology and Other Essays.* Trans. William Lovitt. New York: Harper Torchbooks, 1977.

Hemingway, Ernest. *The Old Man and the Sea.* New York: Simon and Schuster, 1955.

———. *Green Hills of Africa.* London: Arrow Books, 1994.

James, William. *Psychology (Briefer Course).* London: Collier Books, 1962.

Kant, Immanuel. *Critique of Practical Reason.* Trans. Lewis White Beck. Indianapolis, IN: Bobbs-Merrill, 1956.

Kierkegaard, Soren. "Diary of the Seducer." *Either/Or Volume 1.* Trans. David F. Swenson & Lillian Marvin Swanson. Princeton: Princeton University Press, 1959.

Lefort, Claude. "Flesh and Otherness." Galen A. Johnson & Michael B. Smith, ed. *Ontology and Alterity in Merleau-Ponty.* Evanston, IL: Northwestern University Press, 1990.

Locke, John. "An Essay Concerning Human Understanding." Edwin A. Burtt, ed. *The English Philosophers from Bacon to Mill.* New York: The Modern Library, 1939.

Lyotard, Jean-Francois. *The Postmodern Condition: A Report on Knowledge.* Trans. Geoff Bennington & Narian Massumi. Minneapolis: University of Minnesota Press, 1989.

Mea-Wan, Ho. *Genetic Engineering, Dream or Nightmare: The Brave New World of Bad Science and Big Business.* Penang, Malaysia: Third World Network, 1998.

Merleau-Ponty, Maurice. *Phenomenology of Perception.* Trans. Colin Smith. London: Routledge & Kegan Paul, 1962.

———. *In Praise of Philosophy and Other Essays.* Trans. John Wild, et al. Evanston, IL: Northwestern University Press, 1988.

———. *The Primacy of Perception.* Trans. John Wild. Evanston, IL: Northwestern University Press, 1989.

———. *The Structure of Behavior.* Trans. Alden L. Fisher. Pittsburgh, PA: Duquesne University Press, 1998.

Milton, John. *Paradise Lost and Other Poems.* New York: New American Library, 1961.

Proust, Marcel. *Remembrance of Things Past.* Trans. C.K. Scott Moncrieff. New York: Random House, 1934.

Remarque, Erich Maria. *Arch of Triumph.* Trans. Walter Sorrel and Denver Lindley. New York: Ballantine Books, 1998.

Russell, Edmund. *War and Nature: Fighting Humans and Insects with Chemicals from World War I to "Silent Spring."* Cambridge: Cambridge University Press, 2001.

Sartre, Jean-Paul. *Being and Nothingness.* Trans. Hazel E. Barnes. New York: Washington Square Press, 1956.

Shiva, Vandana. *Biopiracy: The Plunder of Nature and Knowledge.* Cambridge, MA: South End Press, 1997.

———. "Diversity and Democracy: Resisting the Global Economy." *Global Dialogue.* Vol. 1. No. 1, Summer 1999.

———. *The Violence of the Green Revolution.* London: Zed Books, 1993.

———. *Stolen Harvest: The Hijacking of the Global Food Supply.* London: Zed Books, 2000.

Stein, Gertrude. *The Autobiography of Alice B. Tolkas.* Middlesex, England: Penguin Books, 1966.

Steingraber, Sandra. *Living Downstream: An Ecologist Looks at Cancer and the Environment.* London: Virago, 1999.

Thoreau, Henry David. *Walden and Civil Disobedience.* New York: Signet, 1960.

Tolstoy, Leo. *Anna Karenina.* Trans. David Magarshak. New York: New American Library, 1961.

Wilson, Edward O. *The Diversity of Life.* New York: W. W. Norton, 1992.

———. "The Bottleneck." *Scientific American.* February 2002.

Wolin, Richard, ed. *The Heidegger Controversy: A Critical Reader.* Cambridge, MA: MIT Press, 1933.

Index

Abram, David, 45
Africa, 76
Agricultural corporations, 76
AIDS, 45
Akhmatova, Anna, 107
Anan, Kofi, 94
Arendt, Hannah, 5, 89
Asia, 76
Athens, 123
Augustine, St., 63

Bach, Johann Sebastian, 81
Beethoven, Ludwig van, 30, 82–84, 92, 103
Being-in-the-world, 2–3, 18, 21, 23, 36, 41, 53, 68–69, 133
Benthamite utilitarianism, 88
Berdyaev, Nikolai, 37
Berkeley, George, 70
Bible, the, 84
Bove, Jose, 76–78, 116
Brutus, Marcus, 138
Buber, Martin, 5, 71, 113
Bush, George W., 99

Caesar, Julius, 138
Camus, Albert, 37
Capitalism, 4–5, 9, 15, 44, 46–48, 57, 74, 87, 89, 97–98, 114
Capitalist corporations, 47, 77, 111
Carson, Rachel, 53–54, 65, 89, 102, 114, 116, 136
    Silent Spring, 53–54, 89, 102, 114, 116, 136
Central America, 136
China, 88
Chomsky, Noam, 136
Colombia, 99
Communism, 74, 97
Consciousness, 3, 18–19, 21–26, 29, 31–32, 40, 71, 91–92
Corporate capitalism, 96, 98, 111

Darwin, Charles, 88–89
Darwinism, 88
Dasein, 2, 53, 71, 133
Descartes, Rene, 3, 29, 35–36, 40, 63, 65
Dillon, M. C., 60
Dostoyevsky, Fyodor, 37
    Crime and Punishment, 37

Du Puy, William Atherton, 101
"The Insects Are Winning," 101

Ecology, 69
Ehrenfels, Christian Freiherr Von, 51–52, 60
Empiricism, 6, 13–16, 21–24, 26–27, 32, 35, 39, 44, 48, 56, 58
Empiricists, 3–5, 14, 19, 27, 35
Environment, 21, 26, 38, 44, 76–77, 102, 112, 116, 121
Environmental ethics, 1
Environmentalists, 2, 25, 44, 67, 111
Esau, 83–84
Europe, 77, 130
Existentialists, 64
Existential thinkers, 24, 36, 91, 133–34, 136

Field of perception, 81–82, 84–87, 93–94, 97–99, 102–3, 106, 108–16, 121, 124, 126–29, 135, 137
France, 130
Francis of Assisi, St., 43
Freedom, 8, 78, 84, 91–98, 127, 136

Gandhi, Mohandas Karamchand, 96, 107
Germans, 98
Germany, 130
Gestalt methods, 51
Gestalt psychologists, 7, 17–18, 20, 51–53, 55–56, 58–61
Gestalt psychology, 7, 15, 51–52, 55–59, 61
Gestalt school, 14
Globalization, 47
Grain (Genetic Resources Action International), 67
Greece, 100, 123
Green Revolution, 76–77
Guatemala, 94–97
Gurwitsch, Aron, 55, 58, 60
Gypsies, 99

Harvard University, 1, 45, 110
Hebrew prophets, 107
Hegel, Georg Willhelm Friedrich, 40, 63
Heidegger, Martin, 2, 5, 26–27, 36–39, 41, 44, 53, 55, 64, 68–69, 71, 116, 133–38
  Being and Time, 68
  "Letter on Humanism," 38
  "The Question Concerning Technology," 5
Hemingway, Ernest, 23–24, 28, 44, 64, 75, 126, 129
  For Whom the Bell Tolls, 64
  Green Hills of Africa, 75, 126, 129
  The Old Man and the Sea, 24, 44, 75, 126
Hitler, Adolf, 98
Hobbes, Thomas, 88–89
  Leviathan, 88–89
Howard, L. O., 101–2, 115
Husserl, Edmund, 16, 39, 51, 55–56, 58, 63

Imperialism, 89
Indonesia, 136
Intentionality, 18, 20, 112–13
Isaac [Bible], 83–84
Israel, 99

Jacob [Bible], 83–84
James, William, 87
Japanese, 102
Jews, 98, 106
Johannesburg, 19

Kant, Immanuel, 5, 40, 66, 68, 78, 122
  Critique of Practical Reason, 66, 78
Kierkegaard, Soren, 6, 37, 64, 91, 105, 112–13
  "Diary of the Seducer," 105
Ku Klux Klan, 106

Lawrence, D. H., 64
  Lady Chatterly's Lover, 64
Leibniz, Gottfried Wilhelm von, 40

Locke, John, 16–17, 29
  "An Essay Concerning Human Understanding," 16
Lyotard, Jean-Francois, 137

Malthusianism, 88
Merleau-Ponty, Maurice, 1–4, 6–9, 13–32, 35, 38–43, 48, 51, 54–56, 58–61, 63–75, 78, 81–89, 91–93, 95–98, 106, 100–104, 106, 108, 117, 121–22, 124–26, 129–30, 133–35, 137, 139–40
  Phenomenology of Perception, 1, 3, 13, 16, 43, 51, 54–55, 57–60, 68, 81–82, 84, 87, 89, 97, 106, 108, 117, 122, 134, 138–39
  The Structure of Behavior, 54
Mexico, 77
Mill, John Stewart, 88–89
Milton, John, 82–84
  Paradise Lost, 83–84
Monsanto, 111
Mozart, Wolfgang Amadeus, 14, 30–32, 52, 55
Muti, Riccardo, 108

Natural environment, 47
Nazi ideology, 99
Nazis, 99
Nazism, 130, 134
Neefe, Christian Gottlob, 92
Neo-Darwinism, 88
Neo-liberalism, 88
Nietzsche, Friedrich, 26–27, 37, 69, 86, 112–13
North America, 77

Occupied territories, 99
Oedipus, King, 43
Ontology, 2, 4, 15, 20, 23, 27, 38–40, 42, 44, 65, 71–72, 88, 91, 127, 133–34

Opium Wars, 88–89
Our Stolen Future, 2

Pacific, 102
Palestinians, 99
Paris, 130–31
Perception, 3–4, 6–8, 14, 17–20, 22–27, 29–36, 38–39, 41–43, 48, 51, 56, 59–60, 64, 68, 70, 78, 81, 84, 88–89, 91–92, 95–97, 100–103, 106, 109–16, 121, 124–29, 135, 137, 139
Phenomenology, 2, 7, 26, 29, 51, 58, 63, 68, 81, 91
Pieres, Maria, 30–32
Plato, 43, 66, 89, 107, 113, 123–24, 136–37
  Gorgias, 137
  Phaedrus 123–34
  Symposium, 123
  Theatetus, 66
Postmodernism, 9, 136
Postmodern writers, 135–38
Proust, Marcel, 32–35, 48, 70, 78, 95–96, 128
  Remembrance of Things Past, 32, 34, 48, 70, 78, 95, 128
Psychologists, 3, 7

Rachmaninov, Sergei, 85
Rationalism, 4, 6, 29, 31–32, 35, 38–39, 42, 47–48, 58
Rationalists, 4, 35, 37, 44
Rebekah [Bible], 83
Remarque, Erich Maria, 130
  Arch of Triumph, 130
Russell, Bertrand, 13
Russell, Edmund, 100–102, 115
  War and Nature, 100

Sartre, Jean-Paul, 5, 92, 134–35
  Being and Nothingness, 92, 134
Seedling, 67
Shiva, Vandana, 46–48, 77, 109–14, 116

Stolen Harvest: The Hijacking of the Global Food Supply, 109
The Violence of the Green Revolution, 117
Sibelius, Jean, 94, 97
Socrates, 43, 66, 100, 123–24
Sophocles, 43, 107
South Africa, 19
South America, 74, 76
Spinoza, Baruch, 40, 63
Stein, Gertrude, 22, 27, 139
Stravinsky, Igor, 108
Suharto, 136

Technology, 5–6, 15–16, 26–27
Third World, 76–77
Thoreau, Henry David, 126, 129
Walden, 126, 129

Tolstoy, Leo, 64, 75, 128–29
Anna Karenina, 64, 128
Twain, Mark, 74
Adventures of Huckleberry Finn, 74

United States, 54, 89, 99, 136

Verdi, Giuseppe, 32, 108
Victorian individualism, 89
Victorian utilitarianism, 88

Wilson, Edward O., 1, 15–16, 45–47, 49, 65, 110–11
The Diversity of Life, 1, 15, 16, 45, 49
World Trade Center, 99
World War II, 100, 102

About the Authors

HAIM GORDON is Senior Lecturer, Department of Education, Ben Gurion University, Israel.

SHLOMIT TAMARI is a graduate student at Ben Gurion University, Israel.